Between Nature and Culture

Global Aesthetic Research

Series editor:

Joseph J. Tanke
Associate Professor, Department of Philosophy, University of Hawaii

The Global Aesthetic Research series publishes cutting-edge research in the field of aesthetics. It contains books that explore the principles at work in our encounters with art and nature; that interrogate the foundations of artistic, literary, and cultural criticism; and that articulate the theory of the discipline's central concepts.

Titles in the Series:

Early Modern Aesthetics
J. Colin McQuillan

Foucault on the Arts and Letters: Perspectives for the 21st Century
Catherine M. Soussloff

Architectural and Urban Reflections after Deleuze and Guattari
Edited by Constantin V. Boundas and Vana Tentokali

Living Off Landscape: Or the Unthought-of in Reason
Francois Jullien, Translated by Pedro Rodriguez

Between Nature and Culture: The Aesthetics of Modified Environments
Emily Brady, Isis Brook, and Jonathan Prior

Reviewing the Past: The Presence of Ruins
Zoltán Somhegyi (forthcoming)

Between Nature and Culture

The Aesthetics of Modified Environments

Emily Brady, Isis Brook,
and Jonathan Prior

ROWMAN &
LITTLEFIELD
INTERNATIONAL
London • New York

Published by Rowman & Littlefield International Ltd
Unit A, Whitacre Mews, 26–34 Stannary Street, London SE11 4AB
www.rowmaninternational.com

Rowman & Littlefield International Ltd. is an affiliate of Rowman & Littlefield
4501 Forbes Boulevard, Suite 200, Lanham, Maryland 20706, USA
With additional offices in Boulder, New York, Toronto (Canada), and Plymouth (UK)
www.rowman.com

British Library Cataloguing in Publication Data

A catalogue record for this book is available from the British Library

ISBN: HB 978-1-78661-075-1
PB 978-1-78661-076-8

Library of Congress Cataloging-in-Publication Data Available

ISBN: 978-1-78661-075-1 (cloth : alk. paper)
ISBN: 978-1-78661-076-8 (pbk. : alk. paper)
ISBN: 978-1-78661-077-5 (electronic)

™ The paper used in this publication meets the minimum requirements of American National Standard for Information Sciences—Permanence of Paper for Printed Library Materials, ANSI/NISO Z39.48–1992.

Printed in the United States of America

To Simon, Warwick, and Ericka

Contents

Acknowledgments

We would like to thank Taylor and Francis, Ltd. (http://www.informaworld.com) for granting permission to publish material, in a revised form, from Brady, Emily. 2006. "The Aesthetics of Agricultural Landscapes and the Human-Nature Relationship." *Ethics, Place and Environment* 9(1): 1–19 and Brady, Emily. 2007. "Aesthetic Regard for Nature in Environmental Art." *Ethics, Place and Environment: Special Issue Environmental Art* 10(3): 287–300.

We are also grateful to Indiana University Press for granting permission to publish material, in a revised form, from Brady, Emily, and Isis Brook. 2003. "Topiary: Ethics and Aesthetics." *Ethics and the Environment* 8(1): 127–42 and Brook, Isis. 2008. "Wildness in the English Garden Tradition: A Reassessment of the Picturesque from Environmental Philosophy." *Ethics and the Environment* 13(1): 105–19.

List of Figures

CHAPTER 2

CHAPTER 3

CHAPTER 4

CHAPTER 5

CHAPTER 6

Chapter 1

Introduction

Environmental Aesthetics between Nature and Culture

Between Nature and Culture explores central issues and problems concerning the aesthetics of modified environments and offers new theoretical insights about the generation and appreciation of objects, landscapes, and environments that are the outcome of complex interactions between (nonhuman) nature and (human) culture, including gardens, agricultural landscapes, ecologically restored places, and land and ecological art works.

There has been growing interest in nonart aesthetics, such as everyday aesthetics, the aesthetics of design and functional objects, and natural and environmental aesthetics. However, within this literature, less attention has been paid to environmental spaces that complicate any simple bifurcation of an assumedly unmodified nature from human artifact. A study of the range of modified environments and our aesthetic responses to them enables a broader understanding of the subject matter of aesthetics, as well as the role of aesthetics in fostering particular human-nature relationships.

This study also reflects, we argue, the varieties of environments that many people most frequently encounter. With ongoing urbanization and rural to urban migration, the natural world is experienced within parks, gardens, urban woodlands, riverscapes, and so on by an ever-increasing number of the world's population.[1] Beyond such spaces, we find suburban gardens and other modified greenspaces, including the agricultural land of rural environments, and the forests, rivers, and wetlands, which humans have been able to restore from degradation by their own hands. Here, we find the farmer working a field or tending to livestock; the weekend gardener weeding his or her flower patch; a family having a picnic in a city park; the walker enjoying a hike through mountains dotted with hill farms; biologists, ecologists, and others working together to restore a canalized river; fishers on the river or the open sea; engineers constructing public works that are situated in nonurban environments; or the environmental artist working in a rural setting. In this respect, the book engages with what are likely to be the more common environmental aesthetic experiences, empirically speaking—that is, experiences of humanly modified environments.

We think that this is important and not just from a position of what constitutes common experiences of the natural world. Environmental philosophers have historically tended to center upon the intrinsic value of natural environments that are perceived to be—rightly or wrongly—untrammeled by humans. It is claimed that a tendency to favor such environments, both in terms of delineating what is of value and the manifestations of such valuing (most notably national and wilderness parks), can lead to a separation between humans and the rest of nature. This tendency leads to preservation of a certain kind of naturalness that is remote and "unsullied" and sometimes hides the violent birthing or creation of such landscapes, most notably the ongoing removal and exclusion of indigenous peoples from land to create national parks (Cronon 1995; Spence 2000).

Recently, environmental philosophy has seen a turn toward pluralistic and practice-based thinking, which focuses on the development of ethical attitudes through regular, sustained, and deliberate human interactions with the nonhuman natural world (Light 2000, 2003; Arntzen and Brady 2008; Marris 2013; Vogel 2015; Thompson 2017). "Environmental pragmatism," in particular, has emerged in response to frustrations among some philosophers that environmental ethics, while making progress theoretically, has been ineffective in influencing policy and creating positive environmental change. The approach is described as

> the inquiry into the specific real-life problems of humanity's relationship with the environment. The new position ranges from arguments for an environmental philosophy informed by the legacy of classical American pragmatist philosophy, to the formulation of a new basis for the reassessment of our practice through a more general pragmatist methodology. (Light and Katz 1996, 2)

In step with this, John O'Neill, Andrew Light, and Alan Holland (2007) consider the role of place and temporal narratives as relevant for shaping ethical attitudes and, in particular, Holland (2012) stresses the significance of meaningful relations between humans and environment for developing an environmental ethic. All three philosophers are interested in moving beyond intrinsic value accounts and toward moral pluralism, with various influences at play. Their inspiration comes from American pragmatism, Aldo Leopold's (1989) Land Ethic and his notion of "land as a community," as well as ideas of flourishing motivated by virtue ethics.

It is this emerging strain within Western approaches to environmental philosophy that we aim to contribute to, but we do so not to denigrate the importance of preserving or conserving less humanized places or indeed the multitude of aesthetic, ecological, and other values and cultural meanings that such places have.[2] Our approach is, rather, one of exploring a new direction in aesthetics through the spaces that are unambiguously the outcome of direct, sustained, and ongoing human-nonhuman interactions.

Finally, reflecting an upsurge of interest in the interactions between aesthetics and ethics, the book also puts forward the central argument that aesthetic experience in both the generation and appreciation of modified environments can contribute in part to developing particular kinds of ethical attitudes toward the natural world. Further, we argue that in the generation of particular kinds of modified environments, aesthetic and ethical values are often intimately bound up with one another, sometimes to the point where the intentional expressions of certain aesthetic values through landscape modification are concurrently expressions of an environmental ethic.

Our overall disciplinary focus is aesthetics and environmental philosophy, though we also draw on ideas from human geography and landscape studies. We should also mention that we take a largely Western approach; although non-Western approaches have a great deal to offer in understanding modified environments, to address this appropriately would require different authors and expertise.[3] This book is jointly conceived and authored and reflects the individual expertise and particular interests that each author brings to the project, while also presenting a diversity of topics to the reader. We hope that our collaborative effort produces an original approach that will be of interest also to disciplines beyond aesthetics and environmental philosophy—indeed, we believe that the study of modified environments requires such an effort.

CONCEPTUALIZING BETWEEN NATURE AND CULTURE

There are a variety of ways in which spaces between nature and culture have been conceptualized and designated. In this book, we work with this variety relatively freely and interchangeably, using concepts such as nature-culture, human-nature, nature and humans, modified nature, cultivated nature, cultural landscapes, humanized environments, human-nature relationships, human-nature relations, and the human and nonhuman. At this early stage in the book, it is useful to briefly set out how we understand some of these concepts, how they have been contested in the literature, and, perhaps more importantly, our approach to the overarching ontological question of what constitutes "between" nature and culture.

Throughout the book, we recognize that there are continuities and differences between the cultural and natural realms. *Homo sapiens* are biological and cultural creatures that share a number of characteristics and basic requirements for life with other organisms. Like other animals, as well as plants, humans need food, water, and light to survive. But humans also create artworks and design buildings and other artifacts, which would seem to set them apart from other species and place them firmly within the realm of culture. Some scientists and social scientists have objected that humans are not the only creatures with culture and that there are cultures of nonhuman animals too, for example, macaques, chimpanzees, and some birds (Heyd 2007, 131). Various kinds of nests as well as other animal homes, such as beaver lodges, have been proposed as evidence of artifacts in more-than-human cultures (Ingold 2000).[4]

Our approach to this complex issue aligns with a widely held position in environmental philosophy that differentiates humans through the technology and particular forms of human agency—ethical and otherwise—that shape their practices. As Robert Elliot (1997, 123) puts it, "Human agency . . . is mediated by a heavy intrusion of culture, social organization and highly structured economic arrangements and is exaggerated by technological capacities." Human agency can be explained through the will, actions, and intentionality, whereas the agency of plants, for example, is best explained through a striving to flourish. If a plant is moved, it will strive to flourish in another place and may live or die. In this way, plant agency may be said to involve a spontaneous kind of being. This agency also results in effects and influences on other things, as when "weeds" push through paving stones, encouraging tidy gardeners to pull them out.

The new geological age of the Anthropocene is the formal recognition by stratigraphers that the nonhuman natural world—or the more-than-human world as some scholars put it—has been pervasively shaped by human cultural forces. The human species has had such a widespread effect on earth systems (including chemical cycles, erosion and sediment transport, etc.) that it is now a geological force detectable in the earth's strata. In light of this new designation, as well as debates within environmental philosophy, it is doubtful that "pristine nature" exists, untouched by direct or indirect human actions. This has led some theorists to question the relevance of the term "nature," to the point where there have been calls to drop the term or replace it (Vogel 1996, 2015; Morton 2007; Lorimer 2015), following on from earlier arguments about human forces ushering in the "end" of nature (McKibben 1989).

Nonetheless, while it may be difficult—and not particularly productive—to support the idea of "pristine nature," we think it is important to resist the collapse of nature into culture and instead hold them in tension with one another. Oftentimes, the types of environments that we discuss in the book are categorized as "cultural landscapes." While we see this category as a useful means to properly foreground the work of human labor, propagation, and territorialization, in shaping the land (including its aesthetic qualities), as well as cultural readings of landscape, it does so in a manner that risks downplaying the role of nonhuman nature.[5] We are thus interested in reflecting upon those environments—farmlands, gardens, land art, mining landscapes—that are quite evidently the outcome of cultural forces, without losing sight of the agency of nonhuman forces.

Indeed, it is important to distinguish our approach from the view that nature is strongly socially constructed—that nature is a matter of human discourse and practice only[6]—while advocating a position sensitive to the many ways that humans interpret and conceptualize nature through a variety of cultural lenses, for example, science, art, literature, oral and written histories, and so on. Our approach, then, is consistent with relatively recent formulations in the social sciences that try to break down—or perhaps, more accurately, *bridge*—nature-culture binaries, such as Sarah Whatmore's writing on human/nonhuman hybridities (Whatmore 2002) and Lesley Instone's "situated entanglements" (Instone 2001, 2004). Our approach, then, is one that seeks to explore the aesthetics of humanly modified environments that inhabit a space between nature and culture—a space that is best understood as a continuum between the two rather than a sharp dichotomy or hierarchical dualism.

What do we mean when we speak of "modified environments"? As we understand them, modified environments are diverse in their character and ontology, ranging from land-based environments with built or cultivated elements to water environments shaped through human management of rivers, the sea, the sky, and atmosphere, as well. Their character will depend upon the role of human causes and the nature of human modification (whether intended or unintended). John Andrew Fisher (2003, 273) makes a useful distinction between "mixed" and "influenced" environments, which helps to unpack the role of humans in shaping environments. "Influenced environments" describe most "natural environments" because they are largely—if not entirely—subject to some kind of human influence or causal effect, whether through pollution, the introduction of exotic species, hunting, historical practices that have shaped the character of a present landscape, and so forth. "Mixed environments" refers to environments that have "artifactual elements—highways, power lines, smoke from distant power plants, jets roaring overhead" (2003, 274). In the book, we use "modified environments" to cover all of these cases—mixed or influenced—though our attention will be focused mainly on cases of intentional human modification of nature, as we see in the design of gardens or creation of artworks in the land. "Modified" is used to identify anthropogenic causes or intentions, keeping in mind that environments are also shaped by ecological and geological processes such as climate, decay, erosion, and animal activity (as mentioned above, beaver lodges, but also bee hives, bird nests, and badger setts).

Modified environments are ontologically complex, having both natural and cultural elements that may or may not be easy to tease apart. There will be obvious cases where the two are distinct, for example, where we find artifacts placed within strongly natural settings such as satellite masts, bridges, roads, pylons, wind turbines, or sculptural artworks in remote places. Here, the artifactual element stands out, sometimes dominating, even marring, the landscape. On the other hand, we also find artifacts overwhelmed by their natural settings, such as an overgrown ruin or an old barn, in which the built feature will have less of a presence. In marine environments, visual evidence can be obvious—a wind turbine or oil platform—or then hidden from surface view, such as a sunken platform. The prominence or lack of prominence of artifactual elements potentially shapes aesthetic appreciation. For instance, wind turbines in a remote, uninhabited landscape can be jarring and seem out of place for some people. For others, the turbines may appear as awesome design objects against a dramatic backdrop (Gray 2012; Saito 2017).

There is also a range of cases that reflect the integration of natural and cultural elements and may be more or less modified. We have already mentioned the kinds of environments that we will be discussing in the book, but additional ones include silviculture, fish farming, canals, reservoirs, and artificial lakes. We do not mean to provide an exhaustive set here; rather, those we have chosen represent a sufficient diversity while allowing an analysis of each with suitable depth. As the book develops, we hope that the ontological complexity of environments between nature and culture will become clear, as well as how that complexity shapes aesthetic appreciation.

Within modified environments, we can often detect the ongoing production of human-nature relationships across time. The contours and variety of such relationships will depend upon different forms of interaction, particular features of environments and settings, as well as particular human intentions, natural processes and different organisms, and so on. But there are a few general points about our approach that can be set out now. We take "relationship" to mean some kind of significant interaction between humans and other parts of the natural world. Just what is significant will depend upon the particular modified environment in question, which should become clear as we proceed. In the book, we argue that a revised and enriched notion of "dialectical relationships" provides a fruitful way to conceive of human-nature relationships.

The concept of the "dialectic" has a long history in philosophy and has featured more recently, for example, in conceptions of nature in Marxist geography and critical theory (Harvey 1996; Vogel 1996; Castree 2003). Within environmental aesthetics, Donald Crawford and, after him, Allen Carlson, use "dialectical relationship" to describe the interaction between humans and nature in relation to some types of environmental art, land art, and gardens (Crawford 1983, 1993; Carlson 2000; Heyd 2006). Crawford writes,

> In a dialectical relationship the two elements of the relationship are conflicting forces whose interaction brings into being some third object—the product of their interaction. The emergent third object may become a new object of aesthetic appreciation, one which results from the synthesis of opposing forces, artifactual and natural. In some cases the synthesis need not negate or dissolve the natural or the artificial; each may retain its identity, and the aesthetic significance of each is dependent upon the interaction between the two—hence the term "dialectical." (Crawford 1993, 144–45)

We are interested in this idea because it captures the meeting and interplay of artifactual and natural processes, with the emergence of a third entity/synthesis expressing the relationship through aesthetic and other qualities.

We would like to revise this idea in several ways, however. First, we think it is important to recognize, explicitly, continuities as well as differences between the cultural and natural, as we have outlined above, rather than conceive of the relationship as grounded in a human-nature hierarchical dualism. Second, we would resist a strong Hegelian dialectic of *opposing* forces and instead place emphasis on dialectical activity as involving an active interrelationship. This approach rests on an understanding of the activity in a dialectical relationship as not necessarily conflicting in a strong sense and instead better described as a kind of tension. In Hegel's terminology, there is conflict, synthesis, and reconciliation, and we take from this the possibility of resolution in terms of positive outcomes for both humans and nature (Maybee 2016). Essentially, we want to avoid conceiving of the relationship as negative or deeply oppositional, while at the same time preserving the dialectical character.

In more concrete terms, consider how through growth and other processes plants respond to but also resist human actions such as transplanting or pruning. In turn, the gardener responds to failure to thrive or grow by providing water, food, and light to plants to enable them to flourish. Plants—within the constraints of a cultivated space—are able to thrive, while the gardener is rewarded with a beautiful, vibrant garden and, perhaps, more broadly, a practice that contributes to the "good life" (Cooper 2006).

One might object that explaining gardens, agriculture, and environmental, land and ecological art, and ecological restoration through the concept of a dialectical relationship is strained because there is no *true* interaction between humans and nature, rather it is really humans acting over or against natural processes. But as Crawford himself points out with respect to land art, much of what we do as human creators depends upon working with nonhuman materials, processes, events, and organisms, and the natural agency they express. We can extend this point to gardens (at least), where natural entities grow and develop in response to specific human actions, such as training or providing more water or light.

The conception of the dialectical relationship developed by Crawford and Carlson is, also, narrower than ours in so far as they make a distinction between a dialectical and a harmonious relationship, where an object or environment resolves itself into one side or the other of the nature-culture relationship. Carlson, for example, argues that French formal gardens and English natural gardens exhibit harmonious rather than dialectical relationships, that is, there is no conflict or strained relationship because art so explicitly serves as the model for nature. In more natural gardens, it is the other way around: nature serves as the model for art, that is, for the design of the garden (Carlson 2000).

In light of this, we make a fourth revision to their idea by adopting a more dynamic notion of the relationship and its synthesis. Rather than conceiving of the interaction as being resolved into one side or the other, we consider the character of each side of the relationship, the ways in which the two interact, and how a synthesis is generated. In most cases, the synthesis itself is not a fixed entity or final product because gardens and other modified environments continue to change or evolve, in line with how the human-nature interactions specific to the place continue. Changes from weather and other forces—even relatively slow geological time—affect, for example, artworks in the land. It is important to keep in mind, too, that while modified environments are the outcome of dialectical relationships, they are also sites of aesthetic appreciation. In this respect, people are responding to and experiencing aesthetic "objects" that are not themselves fixed entities. As the book unfolds, we consider these points in more detail and in relation to a variety of modified environments.

The concept of the dialectic characterizes the ontological entangling of humans and nature in various spaces. It shows how human-nature relations are in a state of becoming, generating new modes of being and new relations that arise out of those new modes of being. Overall, it is able to capture how each side of the relationship mutually shapes the other and moves toward a conception of hybridity rather than dualism.

AESTHETICS AND ETHICS

The environmental spaces that are the subject of this book involve human-nature interactions where humans actively manipulate, fashion, and respond to natural materials and processes, and where nature also shapes and responds to human actions. We will also be exploring some of the ethical questions that arise in these interactions by drawing upon recent work in environmental ethics that focuses on how moral responsibilities flow from human-nature relations.

At times, we consider whether these interactions are grounded in, or are likely to engender, a particular ethical position toward the natural world. Above, we characterized these interactions as dialectical, but from the perspective of environmental ethics we need to reflect upon the distribution of agency within these interactions. Indeed, a dialectical relationship between nature and culture is not, *a priori*, necessarily a mutually beneficial one. Consider, for instance, how a gardener may carefully attend to how plants grow, and seek to work with that growth, or how that gardener may curtail the plants' agency and work against it. This distribution of agency, we argue, is intimately tied to aesthetic values and experiences. In this example, the gardener may be trying to achieve a desired aesthetic quality—say, a neatly arranged garden—that necessitates continued and sustained human intervention.

We do not align positive human-nature relationships with the co-creation of positive aesthetic qualities. Positive aesthetic values may emerge from negative relationships and vice versa. For example, the green lawns that are attractive to many people are often the result of heavy management and use of weed killers that are harmful to plants and wildlife. *Rhododendron ponticum* in the United Kingdom, which is judged to be both "bright" and "attractive," creates shade and biochemicals that inhibit plant and insect life (Brady 2006a, 282; Hall 2015). Inversely, the intentional rescinding of interventionist land management, as seen with rewilding strategies, can create landscapes that are simultaneously judged to be messy or ugly (think here of the sight of dying and decomposing vegetation or the stagnant smells of a blocked river channel) and ecologically valuable for a range of organisms and ecological processes (Prior and Brady 2017). Our approach, then, recognizes the range of aesthetic values—from positive to negative—that are involved in human-nature interactions. For us, this range of values is situated within a common understanding of the concept of "aesthetics" as it is discussed in the literature of philosophical aesthetics. Broadly speaking, the concept describes those properties, qualities, experiences, values, and judgments most closely associated with the sensory/perceptual, imaginative, and emotional dimensions of "objects" and the subjects experiencing them.[7]

Throughout the book, we emphasize the value of harmonious relationships—but as Ronald Hepburn has noted in the context of aesthetics and ethics, "'Harmony' is often so abstract as to possess little informative content" (2001, 55). Our use of the concept opposes harmony with conflict and stresses ways in which humans and nature work together to enable each to flourish, where flourishing is defined as a good. Just what constitutes a harmonious relationship should become clear as we examine the particular contexts of human-nature interactions, for example, in the garden, in the agricultural field, and so on.

BOOK SYNOPSIS

In the next two chapters, we examine the garden as a site of aesthetic interaction between human and nonhuman nature. Chapter 2 explores gardening from a historical, relational, and ethical perspective, using the "picturesque" to frame our discussion. Chapter 3 examines what some might consider to be a rather controversial example of a garden practice: topiary. As we see it, both the picturesque and topiary present challenging case studies against which to examine issues at the intersection of aesthetics and ethics within both creative and appreciative aesthetic practices.

In chapter 4, we consider aesthetic appreciation of agricultural landscapes, contrasting forms of industrialized or intensive agriculture with more traditional or smaller-scale farming, and the landscapes they produce. Chapter 5 explores aesthetic engagement with both environments and nonhuman animals in the creation of environmental, ecological, and land artworks. The main argument we develop in these two chapters is that modified environments are characterized by a dialectical relationship between artifactual and natural processes, which produces both tension and harmony in human and nonhuman interactions. The contours of these interactions indicate some of the ways that aesthetic experience intersects with ethical forms of engagement with modified environments. Chapter 6 investigates the aesthetic values of an ecological restoration project carried out on a former coal colliery in South Wales in the United Kingdom, which culminated in Parc Penallta, a community park. We identify the aesthetic qualities that arise from this particular form of landscape transformation as being "populist," with an overall landscape character of functional beauty.

In the final, concluding chapter, closer consideration is given to the intersections between aesthetics and ethics in the experience of environments between nature and culture. In line with inspiration from recent work in environmental ethics (as discussed above), we provide an alternative to "aesthetic preservationism" and one that is more consistent with the kinds of human-nature interactions that take place beyond remote natural areas.

NOTES

1. We note, for instance, the oft-quoted statistic that over 50 percent of the world's population now lives in urban areas (see United Nations 2015).
2. Our notion of value aligns with Robin Attfield's position: "what there is reason to desire, cherish or foster in virtue of the nature of the state or object concerned." For us, "reason to cherish" can also refer to something having value in its own right (2001, 52).
3. Relevant sources for non-Western approaches will include discussions in the fields of comparative aesthetics, Japanese and Chinese gardens, and sources drawn from landscape design and environmental philosophy in various parts of the world. Within environmental aesthetics more specifically, see, for example, various books and articles by Saito (1992, 1996, 1997, 1998c, 2007, 2017) as well as Miller (1993) and Heyd (2002).
4. Discussions in evolutionary aesthetics have considered whether other species have an aesthetic sensibility or create artworks, but we do not explore that question here (Dutton 2010; Davies 2012; Prum 2017).
5. See especially Carl Sauer's highly influential text *The Morphology of Landscape* (Sauer 1925) and more recently the "new" cultural geography of the 1980s and 1990s, for example, the work of Denis Cosgrove (1998).
6. Such a view has been popular among social scientists but many environmental philosophers reject it, recognizing that while the concept of nature is both culturally shaped and certainly contested, "real" nature does in fact exist on earth (Peterson 1999; Rolston 2002).
7. In the literature, it is common to cite Alexander Baumgarten as the formal founder of the discipline of "aesthetics" in 1735, which may be described as "the science of what is sensed and imagined" (Guyer 2016, 1).

Chapter 2

Gardens I

The Picturesque and Engagement with Nature

In addressing our aim of reaching a greater understanding of modified landscapes and environments, aesthetic appreciation of them, and their place as sites of positive nature-culture co-creation and relationship, human design plays an important role. In gardening, whether designing, creating, maintaining, or just admiring, can we be said to be in a relationship with nature at all? As David Cooper asks, "Is gardening a model of convergence—of a 'fusion' between culture and nature—or is it instead an exercise in human dominion over nature?" (2012, 14). In this chapter, we take a historical journey and consider gardens as sites of response to, or interaction with, nature. We then focus on the aesthetic category of the "picturesque" as an unlikely, but nevertheless helpful, ally in forging a positive dialectic entangling of humans with nature.

The term "landscape" is sometimes connected to an unhelpful idea of nature as either living up to some art-inspired ideal or not, and within that more general critique the notion of the picturesque has been its most derided form. This is understandable as the picturesque is usually interpreted as an admiration of "picture-like," and thus inauthentic, nature. The design of inauthentic landscapes, and even the way we think of nature or place as landscapes, is attributed to this maligned aesthetic tradition. In contrast, we argue an interpretation of the picturesque that is more in accord with the contemporary ideals of appreciating nature's wildness and resisting placelessness. We shall be reassessing the picturesque through the eighteenth-century works of Uvedale Price (a Baronet and owner of an English Estate) and Claude-Henri Watelet (a Parisian essayist and supporter of the arts). We identify six themes in their work (variety, intricacy, engagement, time, chance, and transition) and show that, far from forcing a "picture-like" stereotype on nature, the picturesque guided the way for a new appreciation of wildness—one that resonates with contemporary environmental aesthetics and the appreciation of mixed environments, the value of local distinctiveness, and embodied engagement.

There is undoubtedly a current interest in and love of nature expressed in the form of protecting wilder places as well as traditional rural landscapes. The rise of the appreciation for wild landscapes in the actual environment rather than through the arts was well charted by Marjorie Hope Nicolson in her seminal text *Mountain Gloom and Mountain Glory*. She presents the now almost unthinkable worldview that saw mountains and dark forests as frightful wastelands that were best ignored and hidden from view or used as evidence of the Biblical Fall or Flood—the idea behind the latter being that God conjured up these abominations in response to human sin (1963, 196–98).

It is common to look directly at the concept of the sublime characterized, for example, by the dramatic impact of mountains, for the source of this sea change in attitudes to nature. But we would like to focus on the picturesque in garden design, which is overlooked not least because it is often misunderstood. The picturesque is often characterized both as informally rustic and as following the rules of composition in landscape paintings.[1] The suggestion that one might track the love of wild nature through garden styles and traditions seems immediately problematic, for what could be more unnatural than a garden? The garden has exactly the mix of natural entities, plants, and their constraint or artificial arrangement that is the mark of the unnatural. A piece of architecture is an artifact with no direct pretensions to be nature. A naturalistic style of gardening, by contrast, seems the very essence of the inauthentic; it cannot help but fail to inspire a love of real wild nature. This task will not be easy, but there are aspects and undercurrents here that are worth bringing out.

A BRIEF HISTORICAL SKETCH

The earliest record of garden layouts comes from Egyptian tombs from around 2000 BCE and their typical forms have been confirmed, in terms of what was actually on the ground, by archaeological evidence. A typical design would be a walled area with a central rectangular fish pond with flowering lotus, surrounded by fig trees and edged with flower beds (Winlock 1942). The design is symmetrical, and there are tall shade trees, expressing a basic pattern that crosses many cultures and many centuries. The Persians created elaborate gardens along these lines, and the garden as a *hortus conclusus*—an enclosed space—continued for centuries.

What nature is being excluded varies across the types of nature in question. The Egyptian enclosed garden creates a formalized and ordered place, where the extremes of heat and cold, drought and flood are managed by design. In the English context, the garden was a clearing in the vegetation, a sun trap where overshadowing vegetation is controlled. A crucial motif of the *hortus conclusus* is the exclusion of raw nature.

This exclusion of nature is furthered by the ordering of what is inside the wall. Running through centuries of garden design is the straight axis and the symmetrical arrangement of planting areas and plants themselves. Order and a horizontal axis appear to reign supreme, but these early gardens were also about order and a vertical axis that connected the labor of humans to their God. It was the vertical axis relationship that was important to them, not their relationship to the world outside of the garden wall (Aben and de Wit 1999). In the early botanical gardens, we see an attempt to recreate the Garden of Eden that would, of course, have been thought of as an ordered world (Heyd 2006).

Breaking Down the Wall

For a brief tour of the important shifts in this garden story, we will focus on the various garden technologies that begin to break down the wall. These technologies take away from the garden its role of excluding nature and point to a deep shift in cultural responses to nature.

An early, post-medieval, development was called a mount, a small constructed hillock that would allow those enjoying the garden to view the world beyond its enclosure. Often they were *mounted* by a spiral walk, and the summit graced with a summer house arranged for the best views. For example, at Hampton Court, this was a view of the River Thames.

At the same time as the development of the mount, walls were being pierced with window-type openings to allow for views beyond the garden. Possibly the most innovative pierced wall, because the gap reached to the ground, was a *saut de loup* (wolf's jump). In French formal gardens, this was a means of ending a broad walk with a gap in the wall; the grills of early versions of this form of opening were able to be dispensed with by introducing a ditch and embankment to prevent animals passing into the garden. In a text by A. J. Dezallier d'Argenville of 1709, this is called an Ah Ah! (cited in Fearnley-Whittingstall 2002, 123). This is the precursor to the revolutionary innovation that changed the English landscape on an immense scale, the "ha-ha" (see figure 2.1). Instead of a gap in the wall, the ha-ha, by extending the ditch horizontally, dispenses with the wall entirely. Nature is now so redeemed that it can be enjoyed, not as something glimpsed from the safety of the garden, but as the backdrop to the garden. In fact, the visual illusion of the ha-ha is that the two, garden and nature, have become a seamless whole. The illusion, however, only works because at the same time as the wall is dissolving the garden within it has also been changing.

Figure 2.1. Ha-ha in Gilbert White's garden, Selborne.

These contemporaneous changes could be seen as a way of bringing nature into the garden. Before the break with formal garden styles and the move to the more informal, there was already a trend for setting aside part of a garden for small areas of woodland or informal planting. These were called "wildernesses" (which sounds rather naïve today), and indeed they were often planted with trees in straight rows or with the informality contained within hedges and traversed by straight paths. But there was something going on there, a stirring of some kind of precursor to feelings many people currently have for nature. An early reference to this practice is in Francis Bacon's essay *On Gardening* published in 1625, where he advised having a heath or wilderness of six acres within the garden (Bacon 1950, 49). This was a managed area but managed to look like a flowery heath.

The ha-ha, as we saw earlier, was a way of blending the garden with the land around, and this was spoken of as the garden blending seamlessly with nature. One of the earliest proponents of the style of gardening that the ha-ha facilitates was the English garden designer Stephen Switzer. In his *Iconografia Rustica* (1718), Switzer suggests that opening out the garden allows for enjoying "the extensive charms of nature" (cited in Quest-Ritson 2001, 112).

The garden around the house is further simplified and "naturalized" by getting rid of formal structures and introducing serpentine paths.

These are, in some senses, small steps; there is both a disjunction between what Switzer said and what his patrons allowed. Considered alongside other writings of the time, such as those of the Third Earl of Shaftesbury, we need to be careful about imposing a contemporary understanding of what they meant by "nature." In the early eighteenth century, this could just as well be taken as meaning the beauty of geometrical forms. What we need to do is proceed in small steps rather than grand causal claims, but at least the steps can now take an irregular and not a straight path.

Moreover, another way of looking at the ha-ha is not as an opening of the garden to nature but as taking the gardening mentality of shaping land to specific, art-determined, aesthetic ends and applying it to nature. In that guise, the English landscape garden becomes not a newfound appreciation of nature but a new opportunity for even greater domination.

From this observation, a familiar story is emerging: we can see this leading to the idea of landscapes as scenic, as Italian art-inspired tableaux, and the prominence of English estates shaped to the aesthetic of the landscape paintings of Claude Lorrain and Poussin. This is a well-trodden path and one that leads directly to contemporary critiques, such as Allen Carlson's "Landscape Scenery Model" of inappropriate aesthetic appreciation of nature. Here the prime suspect in distorting our relationship to nature is the picturesque, which he describes as "a mode of appreciation by which the natural world is divided into scenes, each aiming at an ideal dictated by art, especially, landscape painting" (Carlson 2000, 45). However, by looking in detail at the idea of the picturesque, it is possible to see something, or indeed a number of things, going on that could redeem this maligned aesthetic notion and link it to a contemporary feel for human-nature co-creation. In the picturesque, we may also find aspects that fit well with the idea of honoring nature and the particularity of places rather than the urge to "improve" them.

THE PICTURESQUE

First, we need to look in some detail at the commonplace but in fact rather strange definition of the picturesque as being "like a picture." The construction of the word and its usage both in the eighteenth century and since certainly suggests that interpretation, but it is strange nonetheless. It can be argued that this definition is either meaningless or tautological. A landscape is not and indeed cannot be like *any* picture—pictures come in many styles and with many subjects—so, the definition "like a picture" is unhelpful as it means almost nothing. We could extract from it a more focused meaning that goes something like, "a scene or landscape that has the qualities sought by particular landscape painters such that it appears a fitting subject for a landscape painting of that style." But what does style mean here? If it means picturesque, we have a tautology not an explanation. The problematic circularity of the "like a picture" definition is identified by Uvedale Price in 1794 (Price 1842) and discussed very clearly in his objection to William Gilpin's definition. He endorses Gilpin's observations but points out that a definition has to be able to divide some things from others. Quoting Gilpin, as many things "please from some quality of being illustrated in painting" (cited in Price 1842, 78), such a definition does nothing to explain what specifically is the quality of the picturesque. Gilpin traveled to particular types of places, and his observations show an appreciation of particular types of things and not others. As such, it is necessary to examine those types so that a more descriptive definition can emerge (such as we have for the beautiful and the sublime). Here, Price is clearing the way for sharpening up the definition but he doesn't merely make the obvious point that all paintings or subjects of painting are not picturesque—"being a fitting subject for a painting" does not make something picturesque. He has much to say about the role of art in developing our appreciation of landscape; however, he also begins to unravel the idea that our appreciation of landscape is wholly driven by our appreciation of landscape painting. While the weight of historical scholarship on the picturesque is against such a claim (Hunt 1992; Townsend 1997; Pope 2002), it is worth contemplating an alternative. Perhaps we have become so accustomed to the cultural interpretation of taste being driven by art that we forget that at some point someone had to think that there was something about nature that was worth painting. In chapter 3 of *On the Picturesque*, Price presents an alternative origin of the term:

> The Italian pittoresco is, I imagine, of earlier date than either the English or the French word, the latter of which, pittoresque, is clearly taken from it, having no analogy to its own tongue. Pittoresco is derived, not like picturesque, from the thing painted, but from the painter; and this difference is not wholly immaterial. The English word refers to the performance, and the objects most suited to it: the Italian and French words have a reference to the turn of mind common to painters; who, from the constant habit of examining all the peculiar effects and combinations, as well as the general appearance of nature, are struck with numberless circumstances, even where they are incapable of being represented, to which an unpractised eye pays little or no attention. The English word naturally draws the reader's mind towards pictures; and from that partial and confined view of the subject, what is in truth only an illustration of picturesqueness, becomes the foundation of it. (Price 1842, 80)

Nature is then packaged, commodified, and sold as an ideal that can then become a scene that nature either lives up to or not. Of course, the inspiration for painters such as Claude and Poussin was nature as allegorical setting, not nature itself. The early English landscape gardens designed by William Kent or Charles Bridgeman and later those of Lancelot "Capability" Brown were allegorical landscapes; they move out into "nature" as poetic setting as can be seen by the classical temples that they placed as focal points to the scene (Dixon-Hunt 1992, 50–75).

There is certainly evidence in the writings and actions of landscape designers and landowners of the time of exactly this kind of idealized picture making. For example, where ancient trees are felled and even villages moved to create exactly the right vistas at just the right points to be viewed and enjoyed by the trained eye, trained, that is, in reading the landscape as allegory. In one sense, this is the picturesque in full swing, and it has a great deal to do with the emulation of particular styles of painting and a great deal to do with control. But it is not, we contend, the whole picture. To find some redemptive element in the picturesque as a way in to nature, we again return to the writings of Uvedale Price.

Although we often see the picturesque aligned with the break with formality (and it does reject the formal), it is better seen as a rejection of a particular type of contrived informality best exemplified by the landscapes of Lancelot "Capability" Brown and his imitators. Price and Richard Payne Knight, author of the influential 1805 text *Analytical Inquiry into the Principles of Taste*, differ on some quite fundamental aspects (Ross 1998, 121–54), but both abhor the style and prevalence of the Brownian landscape. Their criticisms on this are in accord and attack Brown for being formulaic, boring, and *unnatural*. Price equates Brown and his imitators to quack physicians with a single cure for all ills.

> In both arts the quacks are alike—they have no principles, but only a few nostrums, which they apply indiscriminately to all situations, and all constitutions. Clumps and Belts, pills and drops, are distributed with equal skill. . . . The best improver or physician is he who leaves most to nature—who watches and takes advantage of those indications which she points out when left to exert her own powers; but which, when once destroyed or suppressed by an empiric of either kind, present themselves no more. (Price 1842, 187)

The popularity of Brown's style for a brief but wealthy period in English history meant that many large estates were taking on the same appearance that, though it purported to connect to nature in its serpentine paths and "naturalistic" clumps, bore no relationship to its geographical location. Thus, not only was each estate rendered bland and boring as an individual landscape, but they were all bland and boring in the same way. Copying pictures is not the way to a picturesque landscape, but what is? At the heart of the picturesque is a love of wild nature in a small compass. It is this impulse that might have contributed to current sensibilities and certainly has still more to say. Before mounting a defense of the picturesque as an engaged love of nature-as-experienced, we outline two further technologies that would seem to suggest that such a defense is impossible. These are the viewing station and the Claude glass.

THE STATIONARY VIEW

The term "viewing station" is used to refer to both an identified spot from which one would view a particular scene and a small building placed on such a spot. Contemporary readers of early guides to the English Lake District might be surprised to see how these books sought to shape and manipulate the readers' experience by identifying precisely where one should stand in order to see prescribed views. Compared to the twentieth-century guides of Alfred Wainwright, who charts the whole passage of a walk, these instructions appear rather stilted and closer to the instructions for an amateur landscape artist—which, of course, they were.[2] The buildings are an elaboration of the same idea, as well as a convenience, since they would not only mark the place for the prescribed view but could also shelter the amateur artist while they sketched the perfectly proportioned section of landscape that made a picturesque scene.

These little buildings already seem rather remote from the themes that we want to claim are also central to the picturesque, but the technology impinges even more on any attempt to see the picturesque as not just an art-directed and distanced view of nature. Consider the example of Claife Station above Windermere in the English Lake District. This mock Gothic building (built in the 1790s) was equipped with different colored glass in each of its windows so that the view could be "enhanced" in a way that simulated the atmosphere of each of the seasons: green for spring, yellow for summer, orange for autumn, and blue for winter (Haywood 2006). Other viewing stations, such as one built in 1708 at the Falls of Clyde in New Lanark, Scotland, were equipped with mirrors arranged to reflect the scene. Here the viewers, or amateur artists, would be able to stand with their back to the actual landscape, see its reflection in the mirror, and thus have it contained and easier to reproduce in their sketch book.

Viewing stations described in eighteenth-century guide books and marked on maps are evidence of the growing popularity of wilder landscapes at the time. However, their prescriptive nature and the introduction of buildings to these sites seem to militate against claims for the picturesque as evidencing any kind of authentic engagement with nature. Even so, we need to remember that the idea that these local landscapes had aesthetic qualities was relatively new and, therefore, it is reasonable to suppose that people might have needed some direction in how to see qualities they were unaccustomed to tuning in to.

THROUGH A GLASS DARKLY?

The mirrors of viewing stations find their way into the backpack of the tourist in search of picturesque scenes in the form of the Claude glass. This portable optical instrument was a hand-sized convex mirror tinted gray or sepia. When held to reflect the scene behind the shoulder of the traveler, it could reduce the chosen panorama to picture-like dimensions and coloration. The mirror's tinting was not just to make the scene look like an old painting in the style of Claude Lorrain but also to compress the tonal values and help strip out detail so that the forms and unifying lines of the scene were easier to discriminate and thus reproduce in sketches and watercolors. Gilpin was a keen user and recommender of the Claude glass as a way to assist the recording of picturesque scenes, and these instruments were widely used by travelers following his recommended journeys. What this looks like is exactly the shaping of wildness to art-determined ends rather than an appreciation of what is there. Lars Kiel Bertelsen's treatment of the subject is a standard view. He describes the Claude glass as a sample of "positivistic hardware" and goes on to describe the Claude glass:

> The glass instantaneously positions the operator—for what else should we call the one who handles it?—in a virtual space between the two backs of the case: between the matt black velvet and the ice-cold curved mirror that sends an enfolded slightly distorted perspectivistic projection of the surroundings into the operator's eyes, allowing him/her to read the world like a text. (Bertelsen 2004, 182)

However, even this piece of technology does not need to be read in quite this way. Gilpin is always quick to emphasize that his focus is on observation and on assisting people in their observation of nature (Crawford 1993; Orestano 2003). This suggests that there might be more to the interaction between natural landscapes and the travelers/tourists with their sketchbook and Claude glass than might be obvious to a contemporary reader. As Francesca Orestano points out, the early ridicule of Gilpin (in the figure of Dr. Syntax in a satirical novel by Combe) meant that his work was never fairly assessed. She draws attention to the fact that sketching, as opposed to painting in oils, demanded very little in the way of tools or direct tuition. This meant that it became a means by which the burgeoning middle class, and particularly women, could engage with landscapes and thus "acquire a point of view, maybe even a voice" (Orestano 2003, 171). Gilpin's simple rules about what makes a good picture and his prescriptions about where to find views worth sketching seem terribly restrictive to the modern eye and sensibility more used to engaging aesthetically with nature. Even so, it is arguable that his guidebooks gave many people, who lacked the education of the Grand Tour and access to painting masters, the license to look at nature and the confidence to record it in a way that would be deemed acceptable to others.

One of the things we need to remember about any sketching activity is that the drive to represent, even within the confines of a system, leads to closer observation of what is there. Hence, the eighteenth-century enthusiast of the picturesque, arguably, would have spent more time actually taking in the places they visited than a modern traveler equipped with a camera. And if we ever wonder why the photographs we take fail to capture the real majesty or charm of a place we experienced, this surely has much to do with the indiscriminate nature of the camera lens, which captures all the details rather than conveying the weight or delicacy of the forms that impress upon the human observers the nature of the place they are stopping to contemplate. The Claude glass should not be exalted as an authentic capturing of the real place—it is just that stopping to draw leads to a more engaged looking and perhaps a justification to spend more time closely observing and experiencing a particular place.

The moral of these brief comments regarding the technologies that suggest most starkly that the picturesque should be understood as an art-led, inauthentic engagement with nature is this: even in regard to these technologies, there is more going on than many commentators have recognized. But if this is the case even in regard to technologies that might be regarded as Exhibit A for the prosecution of the picturesque (i.e., as an art-led, inauthentic engagement with nature), then it might be reasonable to think that a more general consideration of the picturesque could reveal that it has done more to inform the contemporary love of nature and engaged appreciation of the specificity or particularity of places than has hitherto been recognized.

THE APPRECIATION OF WILDNESS AS EVIDENCED IN SIX PICTURESQUE THEMES

How exactly does this idea of an appreciation of wild nature come through in writings on the picturesque proper? We now identify and explore six themes, all of which are linked to the current affection for, enjoyment of, impulse to protect, and entanglement with nature.

The first, drawn directly from Uvedale Price, is the idea of *variety* as pleasing (Price 1842, 69). It is a characteristic of nature, in the eyes of this Herefordshire landowner, that it is varied, and that variety is something to treasure rather than obliterate through inappropriate, style-driven, management of the land. Pleasing variety as shown in the changing variety of plants and shapes and forms encountered while walking an "unimproved" country lane is not the only type of variety. The idea of variety as a good in itself finds contemporary resonance with ideas of biodiversity. But, lest we mistake Price for wanting the equivalent of a botanical garden or zoo, remember that in his anti-Brownian polemic we saw that he admires the natural variety that makes one place different from another. Nature in Herefordshire is not like nature in Lancashire, and the garden style that tries to emulate the same form everywhere (particularly one imported from another country entirely) is destroying what Pope, in 1731, had called the *genius loci* (Pope 2002, 207), and what now has come to be called "local distinctiveness" (Clifford and King 1993). We are not claiming that Price, if alive today, would have been taking part in the anti-globalization movement, just that his attack on Brown has resonance with it. For Price, it was each place itself that was the source of rich aesthetic experience. For example, the way a path turned or the way sheep, just through use of a sheltered spot, hollowed out spaces in the banks of a track and revealed the gnarled tree roots. To enjoy and cherish such things was the appreciation of place rather than the hankering for another place. This feeling for *genius loci* and rejection of formulated landscape ideals perhaps has much to say to us today about cherishing nearby nature rather than only hankering for wilderness on a grand "untouched" scale.

The next, connected, theme is that of *intricacy*. Again, it is Price who brings this out explicitly. The term suggests density of detail but Price (1842, 69) adds a specific meaning when he describes intricacy as "that disposition of objects, which, by a partial concealment, excites and enriches curiosity." Variety and intricacy work together, but for Price intricacy is what stimulates our curiosity and imagination. On a country lane, it is this that makes the gnarled tree roots fascinating and drives us on to see what is around the next bend, hidden from view. Though perhaps stretching the point here, this notion of intricacy does seem to point also to an enjoyment of and a desire to know the specifics of nature: to enjoy *this* particular clump of primroses, to want to find out more about *this* particular woodland. We need to remember that the period of the picturesque is also that of Enlightenment empiricism. The rejection of a poetic/emblematic landscape, the new enjoyment and appreciation of a landscape for itself, being intrigued by it, and wanting to understand its intricacies, all fits well with that wider cultural shift. Currently, we can see a similar understanding of nature emerging in something like understanding local ecology, not necessarily from a reductive scientific perspective, but something more akin to an informed amateur naturalism or the resurgence in interest in Goethean observation and understanding nature through his idea of a "delicate empiricism" (Goethe 1998, 307).

The third theme, *engagement*, is developed more fully in the writings of a French essayist Claude-Henri Watelet. In his *Essay on Gardens* (1774), he discusses the problem of likening the work of laying out a garden to that of the work of an architect. He makes the point that with a building we enjoy taking it in at a glance, admiring its organization of, in particular, vertical space. But that mentality applied to the garden has given us the formal structures of straight paths and symmetrical patterns, which do not entice the viewer to even venture out into the garden itself, as Watelet says,

> While an immense parterre or endless allées may astonish, this pleasure lasts but a few moments. One questions the purpose of walking across such great expanses when a single glance has already explored them. . . . But even if one were to undertake this tedious task, this walk that nothing encourages him to hasten or slow down, he would no doubt be like a man who moved his legs without going anywhere. (Watelet 2003, 51)

For Watelet, the garden must be planned out in the place itself, utilizing its natural irregularities and pleasing views, not indoors at a draughtsman's table and not to point to this or that fable or myth but to be itself. The land then becomes something to explore. Note that for Watelet, that exploration is driven by the land itself; it speeds us or slows us according to what is happening in the land. What are the fruits of this kind or engagement? For Watelet, they are psychological, such as peace for the soul, respite from the tiring concerns of society. His accounts of gardens in the essay are given as *perambulations*; they record not just the scenes but also the movement through landscapes and the thoughts and feelings such engagements bring out. For example,

> The last rays of the sun sometimes find me considering in silence the tender concerns of the swallow for its young, or the cunning tricks of the kite attempting to capture its prey. The moon is already up, and I am still sitting. This is an added pleasure. The whispering water, the sound of the leaves in the breeze, the beauty of the sky, all immerse me in a sweet reverie. The whole of nature speaks to my soul as I wander listening. (Watelet 2003, 59)

This has a very contemporary feel: the engagement calls to mind Arnold Berleant's work on the role of engagement in environmental aesthetics (Berleant 1992); the multisensory account chimes with aspects of ecophenomenology (Brown and Toadvine 2003; Brook 2005); and the soul-healing notion of nature contemplation hints at what is later developed in ecopsychology (Winter and Koger 2003).

The fourth theme is *time*. This is obviously connected to movement. If the form of nature appreciation becomes the walk, rather than the picture or isolated scene, it has to take place over time. The shaping of the landscape by the passage of people or animals over time crops up in Price as he eulogizes over the twisting path "shaped by the mere tread of passengers and animals, as unconstrained as the footsteps that formed it" (Price 1842, 71). We are properly carried through the wilder landscape not on the planned paths of the "improver" (a term used for landscape designers such as Brown), whether they be straight or serpentine, but by the paths that emerge through long-term interaction and engagement.

The most obvious aspect of time in the picturesque is the enjoyment of the action of nature over time on the works of humans. The classical temple that had meant one thing, once weathered and decayed, now tells a new story, part of which is evident in the ruin itself and requires no prior learning. Here nature as entropic of human order brings about unplanned changes that render what was once simple and smooth—the epitome of the beautiful—into the epitome of the picturesque. Price gives a wonderful time-paced account of this process.

> Observe the process by which Time, the great author of such changes, converts a beautiful object into a picturesque one: First, by means of weather stains, partial incrustations, mosses, &c. it at the same time takes off from the uniformity of the surface, and of the colour; that is, gives a degree of roughness, and variety of tint. Next, the various accidents of weather loosen the stones themselves; they tumble in irregular masses upon what the perhaps smooth turf or pavement, or nicely-trimmed walks and shrubberies-now mixed and overgrown with wild plants and creepers, that crawl over, and shoot among the fallen ruins. Sedums, wall-flowers, and other vegetables that bear drought, find nourishment in the decayed cement from which the stones have been detached; birds convey their foods into the chinks, and yew, elder, and other berried plants project from the sides; while the ivy mantles over other parts, and crowns the top. The even, regular lines of the doors and windows are broken, and through their ivy-fringed openings is displayed, in a more broken and picturesque manner. (Price 1842, 82)

His account in this passage also ushers in the fifth theme, that of *chance*. What is admired in the variety, intricacy, and developments through the passage of time is that the changes are wrought by chance. The appearance of a decaying ruin can be created, and this was done once it became a popular aesthetic. However, the value that Price and other writers were finding was not in decaying ruins *per se* but in them as emblematic of nature at work, nature indifferent to human likes and dislikes and just doing its own thing. In the English context of a moist temperate climate, this was verdant growth wherever, however, it happened to find a place. Although Price and others talked of creating this effect, it was very much by letting it happen or rather not stopping it happen, such that the "effect" was what would be there if humans were mere participants in rather than conquerors of the land for which we have responsibilities. This opening oneself to chance encounters or developments—going with the flow—and enjoying the surprise of what the other has to offer has been seen today as psychologically healthy. Indeed, the idea of letting nature *be* has become the clarion call of the contemporary environmental movement.

The sixth theme, *transition*, brings us up to date with recent work in environmental aesthetics, but even here we can find useful ideas and precursors in the picturesque. The Brownian design was much criticized by Price, Knight, and later garden designers because it took the green sward—the smooth expanse of grass—right up to the house, having dispensed with the fripperies of intricate parterres and flowering plants. Price does not advise that the roughness and unplanned nature of the picturesque should rule everywhere, despite some early criticisms that suggest he did (Repton, cited in Price 1842, 411–16). Price felt that certain comforts of the home necessitated some order, such as smooth driveways and decorative embellishments close to the house in the garden proper (Price 1842, 267). Hence, rather than writing the informal garden over the wider landscape, he makes a nice distinction between the human environment close to the house as the garden and that which is beyond the garden to be enjoyed for its picturesque qualities. The picturesque can then be seen as a transition between the human realm and the very wild realm of nature further afield as sublime. Indeed, Edmund Burke, writing in 1757, saw it functioning in exactly this way (Burke 1998). Here the picturesque can be seen as a mixing and juxtaposing of the Sublime and the Beautiful such that the smoothness of the beautiful is ruffled with a rustic patina of nature-human relations, and the quasi-threatening disorder of the sublime (untouched wild nature) is toned down to a charming irregularity. Gilpin also brings the two together by describing the picturesque as "beauty lying in the lap of horror" (quoted in Hussey 1983, 22).

What the picturesque allows for is that the transition can be made from one realm to another without us seeing wild places as needing to be toned down and shaped to our sensibilities; rather they are to be experienced for what they are. Also, we can still value the particular qualities that an ordered garden can bring where human design has the upper hand. The transition zone is where both humans and nature have a hand, but neither is tyrant; deer roam, sheep are grazed, trees might be coppiced but this is a more direct working with what nature affords and an enjoyment of its bounty. The idea of the landscape of utility can also be found in Watelet. He develops the idea of the embellished farm where the function of the mill, arable fields, and dairy is not contrived but useful and yet nothing is developed to such a scale that they do not delight the person passing through them with their appositeness to the land and their variety. Here, the picturesque is the landscape that can arise with the human working in and with nature as a participant, striking a chord not just with a Leopoldian-style land ethic (Leopold 1989) but also with recent concerns about the wilderness discourse of early environmentalism leaving no place for a healthy and nature-respecting, but nevertheless human, mode of life (Cronon 1995). The picturesque can help us make aesthetic sense of those mixed communities such as the urban park or sensitively farmed land, as well as bridging the gap between a love of order and love of extreme wildness. Moreover, the picturesque sensibility is an essential part of coming to see the local and specific as a place of experience that needs consideration.

CONCLUSION

We have sought to show that the picturesque, interpreted in this way, has many resonances with understandings of human-nature relationships in contemporary environmental philosophy and the development of an aesthetic appreciation of nature. There is much to poke fun at with its later interpretations and, particularly, its decayed form with its fake ruins and rustic hovels. Indeed, as mentioned above, there were even satirical novels produced at the time (Combe 1878). However, our aim has been to take a closer look before consigning it to the history books of bad design ideas or corrupting influences on understandings of the natural world. There was something interesting about the picturesque that has aided and abetted other cultural ideas to help to see the natural world as both pleasing and as a kind of agency with which we can and indeed must engage.

The implication of this reinterpretation is that we can look to an aesthetic notion other than the sublime to explain and further inspire a love of nature. Moreover, if we think of the sublime as inspiring the love of and drive to protect wilder landscapes, then the picturesque serves a similar role for more local landscapes. In this way, sensitively farmed country and city refuges, such as parkland and even the carefully but not over tended backyard or garden, can be seen not as poor cousins of wild places but as aesthetic places that inspire a love of nature in their own right. The six themes here drawn out of the picturesque can be used to explore our relationship to nearby nature, and we suspect they can do much to honor and help direct our feeling for local nature and endorse care for nature in a small compass. The aspiration to preserve wilderness is not thereby undermined; rather, wilderness takes its place as one type of nature that requires care and respect, but not the only type.

NOTES

1. Recent discussions of the picturesque in aesthetics include Brady (2003), Carlson (2000), Costelloe (2013), Ross (1998), Saito (1998a, 2007), and Townsend (1997).
2. See Palmer and Brady (2007) for a discussion of how the beautiful, picturesque, and sublime play out in Wainwright's walking guides.

Chapter 3

Gardens II

Ethics, Aesthetics, and Topiary

This chapter continues the theme of how the natural and cultural interact in the garden and builds upon the last chapter by considering ethical issues as they arise in the practice of topiary. Topiary presents a fascinating case study for considering these issues because, in many of its forms, the natural growth of plants is highly manipulated by human actions and creates strong modification. We begin by examining the ethical concerns arising from the uneasiness some appreciators might feel when experiencing topiary as an active manipulation—even contortion—of natural processes. Such action pushes against gardening as promoting a harmonious human-nature relationship. We then turn to ways in which topiary might cause an "aesthetic affront" through the humanizing effects of sentimentality and falsification of nature (most often found in representational rather than abstract topiary). Our view is that successful topiary emerges through a dynamic and positive relationship between topiarist and tree, where the gardener works with nature's forms rather than in strong opposition to them. Appreciation of successful topiary, we shall argue, is marked by an experience of an interaction between the tree as a living thing and the role of human artifice in shaping it.

WHAT IS TOPIARY?

Topiary is the practice, or some say, art, of trimming, pruning, and training trees and other plants into ornamental shapes. Topiarists use plants that have growing rates and textures suited to coaxing them into particular shapes and maintaining the shapes over time with pruning. Typical topiary plants are privet and boxwood because of their dense texture and different kinds of yew because of their texture and slow growth rate. Cypress was favored by the earliest topiarists in ancient Rome, and it is used in the present day in warmer climates. One's choice of plant and particular specimen depends on many factors, including what shape one wants to create and how quickly one wants a result. Creation and maintenance of the shapes is achieved through the use of wires, twine, stakes, and, of course, trimming and pruning.

There are different kinds of topiary and many different examples of possible shapes. This chapter focuses on traditional topiary gardens, collections of growing trees that are maintained through trimming and pruning, and trees with some history. Our own experiences of topiary, and many of the examples below, come mainly from one of the largest and oldest topiary gardens, at Levens Hall, just north of Lancaster in England (figure 3.1).[1] The Levens topiary garden was begun by Beaumont in 1689, and many of the original trees still exist.

Topiary also includes knot gardens, espalier (training plants into flat, two-dimensional surfaces), and the architectural topiary of hedges formed into walls, windows, arches, mazes, green rooms, arcades, and columns. Besides these types of gardens, topiary can be found in urban and suburban gardens and, more recently, people have become interested in tabletop or portable topiary, which is created with frames and is more an object for decoration.

Figure 3.1. Topiary Garden, Levens Hall, England.

The fast food of topiary is mock topiary, which is much more like a sculpture in that its materials, glued together branches, moss, and so on, are no longer growing.

We make a distinction in the chapter between abstract and representational topiary. As with this distinction more generally, there is no sharp divide between the two, but there are clear cases of each type. Among abstract topiary, there are the severe geometrical forms of formal gardens—pyramids, squares, and balls—and the more free-form spirals, rings or hoops, cloud-shapes, and judge's wigs. On the representational side, we find, more traditionally, peacocks, jugs, chairs and tables, hunting scenes, fleets of ships, battle scenes, and more recently all sorts of animals including giraffes, teddy bears, trains, and even Disney characters.[2] These days topiary appears in all sorts of forms across the world. Yi-Fu Tuan (2004) reports that a couple of California residents even created topiary hedges in the form of male sex organs, much to the dismay of their neighbors!

ETHICAL CONSIDERATIONS

Before embarking on aesthetic criticisms and potential justifications of topiary, we consider some ethical concerns. There are three broad areas where concerns about topiary as a manipulation of nature could arise: harming living things, harming ourselves, and unfortunate cultural resonances.

Harming Living Things

The first question we might ask is, does topiary harm the tree? The image of a living thing that has been contorted from its natural form or constrained can prompt feelings of disquiet in the observer about how it is for the tree. The idea that it somehow hurts the tree—that the tree is in pain—can be set aside, but there could still be a reasonable concept of harm. The tree cannot feel pain because it has no consciousness or central nervous system.[3] When we attribute pain to the tree, we are expressing sympathy not empathy; we imagine what it would be like for us to be held in that way, to be constricted in our movements and forced to bend to another's will. However, if we do empathize with the tree, that is, if we base our perception of how it is for a tree to be shaped, the picture is less clear. While there is not, in Thomas Nagel's (1974) terminology, "something that it is like to be a tree" and a tree is not, in Tom Regan's (1991) terminology, "the subject of a life," it is still a living thing and as such is goal directed (Hall 2011).[4] Even though, for the tree, there can be no subjective experiencing of being topiaried or having a goal thwarted. The remaining ethical problem is then, whether it is permissible to hinder a tree's goal. We could perhaps invoke a guilt by association move and claim that if topiary is ethically wrong then so is hedging, uprooting ash seedlings between paving stones, and even treating fungal infections, to show that this leads to absurd consequences. However, a critic could still claim that topiary for no purpose other than ornamentation is a separate issue. The only conflicting interest in the topiary case that prevents the tree reaching its goal is the whim of the topiarist, and whims should not override the goal-directed behavior of living things.

The remaining defense here is to point out that a plant's goal-directed behavior is toward thriving, its various systems are geared toward providing its needs, and the intervention of pruning and shaping works alongside these needs. A tree will always be shaped to a certain extent by its environment: the nutrients available, the prevailing wind, and so on; admittedly, it takes human intervention to shape it as a cube, but it can still thrive in that environment. The topiarist's concern is not just for a tree that looks like a cube but one that is still a living, thriving tree. The resulting tree no longer looks like one of its species as depicted in a field guide but then neither would one that germinated next to a wall or under dense tree cover. In managing the environment so that the tree thrives, the topiarist has an important goal of the tree at heart.

Harming Ourselves

A second area of ethical concern is about what effect topiary might have on us as moral beings. This could take two forms: intervening with living things in this way somehow desensitizes us to the harming of other organisms who may be subjects of a life and/or it makes us look at nature as something to be managed always for our own benefit.

As established above, pruning does not hurt a tree, but perhaps overriding the feeling that it does and going ahead with it or admiring its products somehow makes us insensitive to hurting animals. This claim is a little like Kant's (1963) injunction not to be cruel to animals so as to develop the right sensibilities toward humans, except that in this case we are treating something that cannot feel pain as if it could in order to train ourselves to not mistreat organisms that do feel pain. Kant's idea could lead to the right action even if it was for the wrong reasons, although he has been criticized for making this claim with no psychological evidence. However, in the plant/human version, it seems completely unreasonable. Understanding the distinction between organisms that are the subjects of a life, who know their interests and experience pain, and ones that do not is an essential part of knowing how to behave toward both. Clouding the issue with anthropomorphisms does not help us to discover their distinctive needs or treat them well.

Pruning a tree into a pyramid and docking a puppy's tail are both done for appearances; the important distinction to be made is that the puppy is hurt by the action.

Even though we do not learn to treat one thing well by misapplying our sensibilities to something else, the things we see and learn to enjoy could shape our view of the world such that we cannot appreciate different experiences. In this case, the person committed to topiary as an art form in the context of gardening cannot help but see natural environments as "hopelessly messy" and "in need of a good prune." As long as they are not let loose, perhaps this is not a problem! However, increasingly the appreciation of wild nature is seen as morally improving and the human desire to manage and organize nature as a defect of character. A topiary garden as a symbol of human control could thus be seen as emblematic of our errors that have brought about environmental degradation. To want to maintain and continue such practices is surely wrong.

Again, this claim rests on an assumption that people do not make informed distinctions. The person who appreciates a fine example of topiary might be equally able to identify and appreciate a flourishing natural ecosystem. Admittedly, there is a tendency to focus on the things that interest us. Anyone who collects firewood will have experienced the tendency to notice dead branches even when walking in the woods for pleasure; our instrumental interests can intrude and so can our particular aesthetic interests. But that tendency does not mean that we are inherently unable to appreciate different things or likely to see all of nature as malleable. Given the varieties of human-nature interaction within any modified environment, it's important to emphasize that gardens have both aesthetic and ethical significance. Through the agency of the garden, David Littlewood provides a helpful way of explaining the source of that ethical significance:

> Just as the gardener works upon the land, the land and working upon it works upon the gardener. The gardener is one of the things that grows in the garden. The garden also works upon those who do not work upon it, but rather visit and spend some time there. The garden can act, and historically has acted, as a site of transformation. It opens the way for interchange between culture and nature, and does so because it is between nature and culture. (Littlewood n.d., 18)

Interestingly, today the purpose of topiary is often to create a juxtaposition between the formal structure of the garden and the more informal planting of mixed herbaceous areas. The juxtaposition enhances the informality of the other planting, so far from creating a strict order its purpose is to emphasize the mixture of natural forms that are increasingly chosen using ecological principles to establish plant communities.

Historical Injustices

The third area of ethical concern is a complex one that will need some detailed explanation; it could be seen as a deeper look into that second realm of harming ourselves. Possibly the first thing that springs to mind with regard to unfortunate cultural resonances is the social order that prevailed in the period of topiary's heyday. For the English garden, this could be marked by the establishment of the topiary garden at Levens in the seventeenth century. The elaborate gardens of the period were able to be maintained only because of the cheapness of the labor, and the same criticism is made of elaborate gardens of all styles at least until 1914, when everything changed. The informal herbaceous borders of Gertrude Jekyll, the Victorian love of annual bedding plants, and the passion for exotics, all made gardens labor intensive and a luxury to be enjoyed by the rich out of the labor of the poor. We could even go further back to the origins of the term "topiary" in ancient Rome. The *topia* was the ornamental garden of a villa, and the *topiarius* was the slave entrusted specifically with its care (Clevely 1988).

Criticisms of this kind can, we think, be discounted. Contemporary labor-intensive gardens are either managed by reasonably paid gardeners or are the leisure pursuit of their owners or volunteers. In a sense, the exploitation of the past is overwritten by contemporary enthusiasm. Also, a criticism of topiary on this basis would entail a criticism of many revered architectural feats, such as Venice. However, there is a more taxing problem, and it is one that connects to the previous problem of how ordered nature could shape our perceptions. To explain this, we need a somewhat more detailed examination of historical currents. Positioning topiary within a particular worldview, we could see it as emblematic of a particular approach to nature.

To see this clearly, we need to move beyond topiary itself to the landscape design of Italian Renaissance villas. The kind of design principles that dictated the way villas were sited within the topography and the layout of the gardens were developed from an idea of art revealing the hidden order in nature. Far from a simple binary opposition of chaotic nature and human endeavors imposing an order upon the chaos of matter, the idea was to bring out the inherent order that was there but hidden from our view—and the order of nature thus revealed was geometrical. The genesis of the Renaissance designers' ideas was Plotinus (c. 205–70) who saw human reason as the means by which we can both see the reality of nature hidden under the appearance of things and mold it to bring it into a more apparent harmony. In the fifteenth century, this idea of the mathematical interpretation of nature became an artistic concept. The key idea within this was the proportions of the human body as the most perfect embodiment of the harmonies of the universe (Steenbergen and Reh 1996). So, not only was nature to be understood mathematically, that is, through a human construct, but design should use the proportions of the human body as the most perfect expression of that mathematical order. This staggering expression of hubris sits rather uncomfortably with the contemporary consciousness of human destruction of the natural world.

Our claim here is that the unfortunate cultural resonances run symbolically deeper than reflections on the working conditions of the gardeners. It places topiary as quintessentially about human control. This was human control to make the tree not just as we would have liked it to be, but as we thought it *should* be. The design of the garden and the treatment of plants within it could be seen, and certainly were seen, as a project to return nature to its condition before the fall.

We are not suggesting that these historical ideas are foremost in the consciousness of people enjoying topiary, even topiary in an accurate reconstruction of the period, but these gardens were created in part to tell us about who we are and the position of the human being in the order of nature. If they worked at the time, they surely have some subtle effect on us now. Perhaps, we are sensitive to the message behind the medium even when it is transformed and overlaid with later cultural resonances.

Although we have fully addressed some of the ethical qualms about this type of manipulation of living trees, the disquieting thought remains that one could be influenced by a problematic notion of the human role in the "perfection" of nature. This might mean rather subtle influences, and the public understanding of these trends and influences should never be underestimated. For many the historical dimension of garden design, the stories that these features tell are well known and understood as *historical*—people may enjoy the forms and wonder at the hubris in the same glance. However, in moving to questions of the aesthetic appreciation of topiary, we propose that it is necessary to carry forward that disquieting thought and use it as a factor in making judgments about the nature of the aesthetic appreciation and about good and bad topiary.

AESTHETIC CONSIDERATIONS

The shaping and appreciation of topiary presents an interesting problem for aesthetics in the environment. As a gardening practice, topiary brings together human cultural and artificial processes with natural ones, and in many forms of topiary, especially representational forms, artistic intentions and ideas are strongly imposed onto nature. The interaction of these independent components potentially evokes a distinctive kind of aesthetic appreciation, like other kinds of phenomena that lie between nature and art (Carlson 2000; Arntzen and Brady 2008). However, while some people may be captivated and charmed by trees formed into squirrels and top hats, the aesthetic reaction of others may involve a mixture of pleasure and uneasiness. Mara Miller (1993, 85) captures one dimension of the problem when she objects to the "disregard for the plant materials evidenced . . . in the geometric shapes forced upon plants in formal gardens and topiary, where the natural shapes of the vegetation are destroyed to suit the designer's conception, rather than exploited in a dynamic interaction between plan and plant." All of this suggests a set of aesthetic issues connected to potential types of conflict that arise between art and nature or humans and nature through acts of forming and appreciating topiary.

We shall argue that some topiary expresses a positive, dynamic interrelationship between humans and nature and, for this reason at least, we should not dismiss these kinds of gardens outright. As outlined in chapter 1, the concept of a dialectical relationship can help to fill out what we mean by such interrelationships. It is important to adopt a more dynamic approach to human-nature interactions and to recognize that interaction as not being resolved into one side of the relationship or the other. In the case of gardens as a form of synthesis, there is no fixed entity or final product. Gardens continue to change or evolve, in line with how the human-nature interactions specific to the place continue (including how factors such as weather and other environmental changes shape such interactions).

In the dialectical activity involved in the co-creation of topiary and other gardens, we find more or less harmonious and more or less conflicting interrelationships in virtue of the type of interaction that takes place. At times, interactions may be conflicting, as when nature resists human modification or, indeed, when human intentionality is strongly imposed onto nature. This is evident, for example, in the more controlled character of many formal gardens, as Miller (1993, 39) points out, "The French or Dutch formal garden achieves its ends by means of the assertion of control over the living material, revisiting natural growth." At other times, a synthesis emerges in the interaction of humans and nature that reflects a more harmonious relationship, where human activity is more responsive to natural processes, working closely with them in an activity marked by deeper, cooperative engagement.

In some forms of topiary, this relationship is forged through an interaction between the living tree or shrub and the topiarist as sculptor, with the resulting shrub, tree, or hedge conceived of as a co-creation originating in human artifice and natural processes.

Aesthetic Affronts to Nature

We have argued that topiary is not necessarily ethically harmful either to nature or to ourselves, but it is important to distinguish between appropriate and inappropriate aesthetic approaches to topiary. Several writers in environmental aesthetics have sought to capture inappropriate aesthetic treatment of nature (Carlson 1986, 2000; Hepburn 1993, 2001; Saito 1998a). Carlson offers a particularly helpful idea for our purposes in his discussion of an "aesthetic affront" to nature. An aesthetic affront occurs when an aesthetic imposition or indignity is made upon nature. These offences are aesthetic rather than moral because they are "generated by the aesthetic qualities of the object, rather than by, for example, its social, moral, ecological, or other such qualities" (1986, 637). Carlson is mainly concerned with aesthetic affronts caused by some works of environmental art. For example, as we shall discuss in chapter 5, works of environmental art by Michael Heizer alter the landscape by cutting grand forms into it, thereby forcibly imposing human actions onto nature.[5] However, not all environmental art is open to this type of criticism because the nature of the artwork itself, its sensitivity to its natural/environmental setting or site, as well as the intentions of the artist, may all be factors that determine whether or not it is an aesthetic affront (Crawford 1983; Carlson 1986, 2000; Ross 1993).

There is a sense in which topiary imposes artifactuality upon nature (and also a sense in which it might be conceived as a type of environmental art).[6] But it is not the case that every imposition of artifactuality onto nature will cause an aesthetic affront. As we have suggested, topiary is potentially a form of gardening that works dynamically with nature, moving closer to it, rather than away from it and against it. Also, topiary works within a particular context—the garden—where artifactuality is already present in our practices.

However, there remain some topiary practices that could be seen to cause an aesthetic affront. One problematic practice lies in the particular shapes and forms that the gardener fashions. Some representational topiary, such as topiary animals, set out to imitate animals and other shapes, which in many cases manipulates nature for the sole purpose of entertaining humans (figure 3.2). These forms are much further away from nature's own forms than some abstract topiary, such as clouds and imperfect, less severe shapes. The aim of some representational topiary is not to work dynamically with nature's forms, or to draw attention to them, but rather to point beyond nature to a human-centered world.[7]

Although representation is a common aim in art, gardens constitute a practice that involves working directly with nature. Some garden practices place value on nature only insofar as it meets a standard of art, as we saw this in *some* ideas of the picturesque, but others do not, and these may adopt an appropriate attitude toward nature in terms of the aesthetic choices made by the gardener. We believe that most representational topiary does not encourage such an attitude and, instead, constitutes the *imposition* of a human view onto nature through an aesthetic means.[8]

Figure 3.2. Frog Topiary. Contribution from the city of Atlanta, Mosaïcultures Internationales de Montréal, 2013.

Trivializing and Sentimentalizing Nature

The humanizing tendency of some representational topiary leads to a falsification of nature. Ronald Hepburn (1993, 69) has offered some interesting ideas on this, pointing to a trivializing approach as one that "distorts, ignores, suppresses truth about [nature's] . . . objects, feels and thinks about them in ways that falsify how nature really is." Although his examples are directed at less humanized environments, we can apply his ideas here to address the relationship between nature and humans in the garden. There are different ways in which trivializing might occur and understanding them helps to identify the problem with representational topiary.

Hepburn (1993, 70) argues that sentimentalizing nature falsifies nature because it "falsely posits human feelings and human attitudes in the non-human—or more likely posits failed human life and human attitudes instead of successfully attained non-human life." Anthropomorphism is usually discussed in the context of projecting human ways of being—emotions, experiences—onto other living species, mainly other mammals, but it has an interesting application to representational topiary. We are made to see a natural organism as another form, a thing that it is not, by projecting another form onto it. This sentimentalizing takes place also when we take a too limited or narrow view of nature and may take the form of appreciating nature only from one point of view or by misperceiving it altogether. Fantastic representational topiary is a case of this because it fails to capture the natural aspect of the tree in any appropriate sense, that is, in a way that is responsive to the tree as a growing, thriving thing. It imposes one way, a fanciful human way, of conceiving of a tree. (This is not to deny that nature can be fanciful in at least a metaphorical sense; but then nature does not impose its "fancifulness" on other things.)

Not all anthropomorphic responses are bad, since they can identify similarities, say, between humans and other mammals, but we do need to be critically aware when we respond in this way to nature. Gardens are cultural landscapes insofar as they involve human purpose and intentionality, but some representational topiary goes too far, erasing the tree or shrub as a living thing. It is much further away from the truth of the tree than other forms of topiary. We might say that in the way that uncritical anthropomorphizing tries to make nonhuman living things into human kinds of things, some representational topiary tries to make trees into art. We think the aim of topiary ought to be to bring out the interrelationship between humans and nature, not simply to create an artifact on the one hand, or allow nature to take its own course, on the other. Making topiary into an artifact commits a category mistake in the garden setting.

The category of kitsch arises too in the context of representational topiary. Kitsch has been defined as two opposing or very unlike things or qualities brought together in a single medium. It has also been criticized by Gordon Bearn (1998, 66) "for being too easy, for being too formulaic, and for being a lie." Objects we may be inclined to call kitsch, "Muzak, Eiffel tower pepper grinders, ice cubes shaped like breasts," and so on, may be enjoyed by some, but in art criticism, kitsch is usually defined as a negative aesthetic category, and in bad taste. Examples of kitsch topiary might be Disney characters or trains, where a tree is made into something entirely different from itself, and where two mediums come together inappropriately, causing a deep aesthetic conflict. This conflict arises because the topiarist, in the aesthetic choices that are made, ignores the natural qualities of the tree and uses it only as sculpting material, rather than trying to work with the tree and respond to it as a natural entity. Working toward rather than against positive interaction between topiarist and tree may have important implications for appreciation. Aesthetically good topiary enables the viewer to see and appreciate the tree and its own qualities, in contrast to bad topiary that negates or hides them.

Imagination and Whimsy

In appreciating gardens, we understand the role of human creativity in them, and we appreciate it as part of the complexity of their value. Along with this comes the valuing of imagination in the garden. Imagination most often shows itself in garden design, for example, how gardeners seek to evoke imagination through "invitation," in the way they design paths, prospects, and by presenting us with phenomena and designs to explore through imagination and the senses (Ross 1998).

Topiarists also use imagination as they prune and sculpt trees. Representational topiary represents a strong use of imagination in so far as it tries to get us to see a tree as a train, a squirrel, or whatever. Imagination can, however, spread its wings too freely. In the context of the artworld, we encourage creative imagination to spread its wings, but when humans are working more directly with nature, as in environmental art or gardens, imagination ought to be directed in ways that work positively with nature's forms.

Whimsy is an aesthetic category that points to qualities of playfulness, joy, and humor. This quality can be found in art works but also in gardens and especially in topiary. Whimsical qualities are not necessarily objectionable, but *overly* whimsical representational topiary overly humanizes nature, making gardens into theme parks and upsetting the harmony that gardens create between humans and nature. Topiary squirrels and trains are too whimsical. They are in bad taste, and as such they also constitute a kind of aesthetic disrespect for nature. There may, however, be cases of abstract topiary that are whimsical without being in bad taste. Here, we have in mind a particular bush at Levens, a small, abstract dome shape with a short, delicate trunk growing up, topped by a tiny ball of leaves.[9] The gardener at Levens improvised the tiny ball shape from the trunk that kept growing up. Rather than cut it off, he responded to the pattern of growth and shaped a second element on the top of the bush. The dome shape alone would be a typical abstract topiary bush, but with the addition of the tiny ball on top, the effect is light and playful.

TOPIARY AND POSITIVE RELATIONSHIPS BETWEEN HUMANS AND NATURE

Now, we would like to bring some of these criticisms together to help us to distinguish between successful and unsuccessful topiary within the aesthetic context of topiary practice and appreciation. This is not an easy distinction to make. First, it helps to experience the cases we discuss firsthand, or at least to view a picture of each in order to understand why they may be in good or bad taste. Second, the distinctions are not always clear-cut, and not all cases are straightforwardly categorized as problematic or unproblematic. The best course is a case-by-case approach, but we also try to show how, generally, abstract topiary is less objectionable on aesthetic grounds.

Problematic representational topiary includes cases where the aesthetic aims and actions of the topiarist create a human imposition on nature or trivialize nature. These are frivolous pieces that move in only one direction in respect of the relationship between humans and nature. Trains, Disney characters, teddy bears, and suchlike move toward artifactuality and intentionality rather than revealing or encouraging an interrelationship between topiarist and natural processes. Some topiary that falls into a gray area between the abstract and the representational may constitute successful cases, especially cases of imperfect forms. Here, we have in mind the "judge's wig" topiary and the "top hat" topiary from Levens Hall. These are traditional topiary forms in which one can identify a representational shape, but they are abstract enough to express more of the tree's shape than other more explicitly representational forms.

Within the category of abstract topiary, we find objectionable attempts to impose onto nature aesthetic qualities found in perfect geometrical forms rather than to work more with nature in an interplay that expresses a dynamic relationship. Examples of problematic abstract topiary include the rigid symmetrical forms of conventional topiary such as pyramids, cubes, and so on, commonly found in formal gardens. Successful abstract topiary is wide-ranging, and it is typified by the topiarist working with nature's forms, allowing the tree to play a central role, where the individual improvises upon natural material. In some cases, these forms have their origins in outgrown geometrical shapes. For example, some of the abstract topiary trees in the Levens Hall garden, which are very old, have taken on a life of their own. The result is less severe shapes that the gardener responds to in his choices about how to prune these trees. This kind of topiary feels much less contrived and artificial, although it is still a product of the garden. Examples here include the following: ring topiary, a series of rings or hoops formed by yew branches, one on top of each other, revealing the complex intertwining of branches within the tree; clouds, great abstract masses of privet or yew lying close to the ground and forming various asymmetrical rounded shapes, reminiscent of clouds; perhaps the small Levens' dome with the tiny ball top; and other imperfect or less ordered abstract forms.

In respect of abstract topiary, more conventional topiarists might be able to learn from contemporary garden movements, while at the same time maintaining their styles and tradition. Paul Cooper (2001) discusses various forms of "living sculpture" and the gardens of Pearl Fryar. Fryer had no previous experience of topiary and taught himself how to sculpt trees. His style could be described as inventive but free-form, allowing the existing shape of the tree to direct his work: "Pearl Fryar's topiary has been described as 'botanical sculpture,' and his abstract creations certainly bear little resemblance to mainstream work. Sweeping curves, arches, and spirals lend them a dynamism rarely seen in conventional topiary" (Cooper 2001, 26).[10] Indeed, it might be difficult to tell the difference between some of Fryar's topiary and natural trees themselves, if it weren't for the setting of the garden.

We are not saying that good topiary is necessarily topiary that looks like nature. Topiary is a garden practice that involves humans shaping and modifying nature in explicit ways. However, it is possible to create topiary that is in good taste in so far as it seeks to express some distinctive qualities of natural forms.

THE AESTHETICS OF IMPERFECTION

We would like to conclude by suggesting a useful way to approach topiary as an aesthetic category. The aesthetics of topiary could be articulated as an "aesthetics of imperfection," where imperfection is viewed as a positive aesthetic quality. This category has been used to describe the Japanese aesthetic taste for imperfection, humbleness, and natural forces, such as weathering and organic qualities, as opposed to showiness and flashiness (Saito 1997).[11] Another take on the aesthetics of imperfection has been put forward in the context of jazz music (Hamilton 2000). Here, imperfection is associated with the freedom, creativity, and verve of improvisation. Most jazz improvisation works with a standard jazz tune and plays freely with it to bring out its qualities and form.

In the context of topiary, we might think of it as an "art" of imperfection, where humans work with trees to create an object or environment that is expressive of something more dynamic, a creative relationship between culture and nature. Our favorite ring topiary yew tree at Levens Hall illustrates this type of aesthetic (figure 3.3). The head gardener, who regularly prunes and sculpts this tree, works with its natural form, improvising with nature's shapes, while his work is still directed very much by the organic processes of the tree as it continues to grow. The tree is two or three hundred years old, and it leans over from the weight of its age. The gardener is responsive and works with this shape rather than seeking to correct it or force it into a more perfect form. As we see it, the best trajectory for traditional topiary practice would be to evolve by moving closer to nature, that is, to work more with natural processes, in the spirit of co-creation, rather than harshly imposing human order or human ideas onto it.[12]

Figure 3.3. An example of pleasing, abstract topiary, currently tended by head gardener, Chris Crowder, at Levens Hall.

NOTES

1. For additional images of topiary at Levens Hall, see http://www.levenshall.co.uk/levens-hall-gardens-historic-topiary-in-cumbria/about-the-gardens/the-topiary-gardens.html. Accessed June 27, 2018.
2. Probably the most famous collection of animal topiary can be found at Green Animals Topiary Garden in Rhode Island. This garden, begun in the eighteenth century by Thomas Brayton, has over eighty topiary animals.
3. The nascent field of plant neurobiology has sought to establish, for example, that plants communicate, but there is no evidence of states of pain experiences.
4. For example, Hall (2011, 158) argues that plants are "communicative, relational beings—beings that influence and are influenced by their environment. . . . Plants display intelligent behaviour in order to maximize both their growth and the production of offspring." He does not argue that they are fully "persons" in the philosophical sense.
5. Crawford (1983, 51–52) cites some of Heizer's works as examples of a third type of dynamic relationship between nature and art, but one in which one side of the relationship dominates and even destroys the other.
6. Stephanie Ross (1998) describes environmental art as akin to avant-garde gardening. The idea that gardens can be classed as works of art is explored in depth in Miller (1993).
7. Of course, animal topiary represents natural organisms, but our point is concerned more with the topiarist's relationship with the tree and its natural aspects.
8. David Cooper (2006, 103) largely agrees with our position in this chapter, although he writes that some topiary is not only an aesthetic affront, but also "instantiates a failure to exercise certain virtues—of care and humility, say—induced by garden-practices engaged in with reflective understanding."
9. Perhaps one could say that overly whimsical topiary *lacks* imagination in so far as appropriate imagination would be that which is guided by the aim of shaping topiary that reflects nature's forms rather than going against them in such a strong way.
10. Some of Fryar's topiary is representational, so our point about his style would not necessarily apply to those cases.
11. There are other aspects of Japanese aesthetics that are helpful in supporting our position concerning topiary as a gardening practice, such as the view that Japanese gardens seek to reveal the essential qualities of natural phenomena. However, there has been some debate about this apparent respect for nature in the garden setting, and it is too complex for us to be able to address here. See Saito (1992, 1996, 2007).
12. In writing this chapter, we are grateful to Chris Crowder, head gardener at Levens Hall, for his practical knowledge and philosophical reflections on topiary practice.

Chapter 4

Agricultural Landscapes

In this chapter, we move on to explore what might be considered to be gardening on a vast scale and gardening that is productive of food: agriculture. Land-based agriculture comes in many different forms, from small-scale to large-scale, organic, biodynamic, "natural" farming, industrial and intensive, fields of wheat, fruit orchards, rice paddies, and so on. To explore the human-nature dialectical relationship that emerges in this context, we focus on interactions with the land in the generation and aesthetic experience of intensive (or industrial) farming in contrast to three traditional agricultural practices, hedge-laying, coppicing, and stone-walling. We shall argue that more human-nature interactions, and greater aesthetic value, may be located within "traditional" agricultural landscapes, that is, landscapes that involve closer interactions with the land, and which are more often associated with smaller-scale and agrarian types of farming.

SITUATING AESTHETIC ENGAGEMENT WITHIN AGRICULTURAL LANDSCAPES

Aesthetic engagement with traditional agricultural landscapes is situated within modes of inhabiting the land. Importantly, it is not characterized by observing grand views from a distance, which has become a conventional way of enjoying wilder landscapes or "natural scenery." As we shall characterize it, the aesthetic response to agricultural landscapes is importantly embedded within human practices that involve a direct and close relation between individual, community, and natural forces. The relationship that emerges between humans and nature is often expressed through hybrid nature-culture characteristics in the land itself.

Aesthetic is here understood richly and critically, not only to include visual and surface qualities but also to reach significantly beyond them to bodily and sensuous experience, imagination, emotion, and knowledge (broadly understood). Some forms of traditional agriculture, such as hill farming, may involve closer interactions with the land and its creatures, characterized by an aesthetics of embodiment, where the individual is immersed in environment. The aesthetic response here is more situated in so far as the farmer is engaged with environment rather than separated or distanced from it (Berleant 1997, 2004; von Bonsdorff 2005).

The concept of the aesthetic is commonly understood to be distinct from the functional or utilitarian, except in architecture. The arts and natural beauty are perceived as lying within a sphere distant from our practical, everyday lives for the reason that their aims are not practical (Kant 2000; Parsons and Carlson 2008). Rather than being opposed to the functional, aesthetics within agricultural landscapes sits alongside or is integrated within practical, productive activities that are not ordinarily or mainly aimed at an aesthetic effect. Indeed, the aesthetic response is in some cases an outcome of working the land, especially in the way that working the land brings about a deeper engagement with natural processes and qualities, leading to a richer and more complex appreciation than a superficial encounter.

We have been using the notion of dialectical relationship in the book to explain the relational, dynamic character of interactions between nature and humans. It captures differences, conflicts, and harmonies, and how either side of the relationship is not necessarily subsumed by the other. The dialectic captures the ontology of humans and nature as entangled, and in concrete terms, gardens, parks, agriculture, and other modified landscapes can be said to express a synthesis of human-nature relations. That synthesis is not a fixed state; rather, it ought to be understood as a state of becoming, generating new modes of being and new relations that arise out of those new modes of being. The dialectic provides an alternative to dualistic thinking with regard to human-nature interactions and shows how each conditions the other.

Before bringing this notion to bear on agricultural landscapes, a few critical clarifications are needed. Generally, it ought to be recognized that there are a variety of activities in which humans modify nature in ways that are considered acceptable, given certain aims. Apart from the most obvious cases (open cast mines, forms of pollution), human modification of natural environments may or may not constitute environmental harm, for example, managing the land to lesser and greater degrees, creating various types of gardens, parks, and so on. Whether such modifications are considered harmful will be determined largely on a case-by-case basis and according to one's environmental ethics. The more biocentric a view, perhaps the less inclined one is toward finding such modifications acceptable. While some practices are problematic, others may be seen as genuine cases of humans interacting with nature in ways that generate aesthetic qualities that embody positive value through the harmonious interplay of artifactual intentionality and natural processes. In any case, we are aware that some human-nature interactions will involve resistance on both sides of the relationship and potential conflict.

Resistance and conflict are also relevant to thinking about how humans are affected by natural processes that are more spontaneous and less predictable than human actions. Natural processes are dynamic from moment to moment and age to age. The forces of meteorological and geological change create conditions sometimes favorable but often extremely difficult for human habitation. Agricultural landscapes are deeply shaped by these and other natural processes. Nature becomes a measure in relation to which our abilities may be challenged or nurtured. Good or bad weather determines the success of the harvest or the quality of grazing. Indeed, it is because natural processes resist our attempts to control them that humans develop various ways and means to live in the land, sometimes harmoniously and sometimes not. In the setting of rural, farmed landscapes, nature is by no means mute or subservient. It is most often a partner, willing or unwilling, in the co-creation of hybrid environments.

Finally, these reflections point to the interesting and apparently inconsistent ways natural processes may be judged in terms of their effects on the environments and objects of dialectical relationships. For example, positive aesthetic value is often placed on objects with a weathered appearance, especially if those objects are fashioned from natural materials, such as wood or stone. An old wooden fence enclosing a pasture may be considered more beautiful than one newly constructed, perhaps because nature's effects have made the fence look more natural, more in keeping with its rural environment, the fresh evidence of human making having been worn away. To a farmer, however, the fence may be unappealing, having lost some of its strength and usefulness. Ruins present another case of a dialectical relationship where an artifact's character is significantly shaped by natural processes (Crawford 1983). As discussed in chapter 2, the picturesque favored ruins for the way the effects of decay, weathering, and plant growth evoked the passage of time and the mystique of the past, with fake ruins being created in landscapes to satisfy this taste. Beyond the picturesque, ruins of all kinds continue to be appreciated for their aesthetic qualities, either as architectural remains alone or as presenting an interplay of cultural artifact and natural processes (Scarbrough 2014).

On the other side, things made from more artificial materials such as plastic and concrete are seemingly made less appealing as they become worn in various ways. For example, weather-stained concrete or an old plastic chair will appear ugly compared to how these materials appear when nearly perfect. For materials that are located more in the realm of the artifactual, the effects of weathering may be seen as a type of damage because it is in conflict with the finished character of artifacts intended to resist natural forces. An exception to this lies in the aesthetics of imperfection associated with Japanese aesthetics (as we discussed in chapter 3), where the modesty of imperfection is valued more highly, as in the tarnished rather than shiny surface of a teapot (Saito 1997, 2007).

THE HUMAN-NATURE RELATIONSHIP AND INDUSTRIAL AGRICULTURE

One might ask where agriculture fits into environmental aesthetics, since the landscapes of both traditional and industrial farming are principally functional. Their design and appearance are a means to an end: the production of crops or raising livestock. At the same time, these environments may have great aesthetic appeal, especially the pastoral landscapes associated with more traditional and small-scale farming practices. Compared to gardens, where one aim among others is to create an environment of aesthetic interest, the beauty of farmlands is for the most part unintended. Some environmental artists have been interested in working with crops to create artworks in the land, but in these cases there is a clear artistic intention to use the medium of living, cultivated materials to create an impermanent, organic work of art (Cooper 2001).

This is not to say that agricultural landscapes ignore aesthetic considerations altogether. The care taken in the upkeep of farm buildings and keeping the area free from waste or rubbish may be for both aesthetic and utilitarian reasons. Moreover, taste and nostalgia for the pastoral encourage the maintenance and conservation of traditional farmsteads for aesthetic reasons as much as for the desire to preserve a certain way of life. Indeed, the aesthetic qualities of these farmlands—rolling green pastures, white clapboard farmhouses, and red barns—carry meanings and associations of an appealing way of life few have access to these days (Schauman 1998; Egoz et al. 2001).

In a comparison of traditional agriculture and the new landscapes of industrial agriculture, Carlson shows how we may be able to come to aesthetically appreciate the new landscapes, despite the fact that to the "untrained eye" they appear bland and monotonous with their "large flat uniform fields" and "featureless metal sheds" (Carlson 2000, 182). In aesthetic appreciation, we also ought to be able to overcome the dismal social consequences of the decline of rural communities and depopulation that has been caused by agribusiness. Instead, we can develop an eye for the "breathtaking formal beauty: great checkerboard squares of green and gold, vast rectangles of infinitesimally different shades of gray" (Carlson 2000, 185). Carlson is not saying that industrial agriculture has more aesthetic value than traditional agriculture; his aim is to try to show that positive aesthetic value may be found in the new, aesthetically challenging landscapes. Nonetheless, his case is controversial and not easy to defend, even if one tries to be open-minded. There may be aesthetic merit in the new agriculture, and perhaps we ought to attempt to change our tastes, but there are good aesthetic (and moral) reasons for favoring traditional methods of farming (Brook 2013, 116).

Carlson argues that because agriculture is a functional landscape, its appropriate aesthetic appreciation must take into consideration function and necessity. The aesthetic merit of industrial farmlands is importantly linked to the fact that they function well and are "extremely necessary" for food production. In agriculture, years of trial and error together with the pressures for production have resulted in landscapes that can be appreciated as paradigms of good design—crisp, clean, and uncluttered in appearance and expressive of ingenuity, efficiency, and economy. Moreover, not only the vast open fields and the orderly farmsteads, but also both the machines and the buildings that occupy them, are aesthetically richer when appreciated in terms of how well they perform their functions (Carlson 2000, 187–88).

On a superficial level, this may be true: there is no question that the landscapes of agribusiness produce a great deal of food "efficiently." But more critical reflection on these assumptions throws up a number of serious objections to industrial farming practices (Kimbrell 2002). In a nutshell, this large-scale, commercial agriculture does not fulfill its function well because it is unsustainable. Carlson recognizes the harm underlying these landscapes in terms of the pastoral way of life they have superseded, but he does not take sufficient account of the environmental problems associated with them, such as damage from fertilizers and pesticides, erosion, soil compaction, unsustainable use of water resources, the decrease of wild animal populations due to destruction of habitat, and so on (Pollan 2006; Shiva 2016).

If appropriate aesthetic appreciation takes place through an understanding of the landscape's function, or, as Carlson puts it in an earlier section of the essay, knowledge relating to the questions "what does it do and why does it do it? . . . we must know the different uses to which the land has been put and which have thus shaped its look" (Carlson 2000, 185), then it seems we are obliged to seriously consider information about not only positive effects but also negative effects in relation to its function. In other words, Carlson's own argument requires consideration of what underlies—both good and bad—the pleasing aesthetic qualities that he points to.

Once we take on board information about the problems in the new agricultural landscapes, the case is weakened for developing an aesthetic appreciation of them. These landscapes embody a deeply conflicting dialectical relationship between humans and nature in the way that human technology

Figure 4.1. Intensive agriculture in Scotland. East Lothian from the air.

is so strongly imposed on nature, especially when compared to other forms of agriculture. Their aesthetic qualities are connected to and partly generated by practices that harm the environment, for example, the fertilizers and pesticides that keep the fields and their produce unnaturally perfect in appearance or the gleaming silver sheds that conceal chickens and pigs raised in inhumane conditions (see figure 4.1).

It might be conceded that the "efficiency" of industrial agriculture means that many areas of wilder nature that are so enjoyed are preserved because there is no need to farm them. The beautiful forests of the eastern United States might otherwise have been deforested in the interests of agriculture had it not been for agribusiness being located in other parts of the country. Essentially, the claim is that much natural beauty is purchased by large-scale agriculture. This presents a difficult conflict between aesthetics and ethics, but the claim is not in itself a reason to value or justify the harmful practices of industrial agriculture or continue such practices in the future. Moving away from those practices need not present a stark choice between preserving designated wilderness areas, say, and intensively farming other parts of the environment. Proponents of small-scale and organic farming aim for sustainable agriculture that encourages biodiversity and, generally, brings humans closer to the land (including connecting consumers more directly with the production of the food they eat). This type of farming presents a viable middle ground and one which, in many instances, will be in keeping with the aims of environmental conservation and natural beauty.[1] In any case, we suspect that the objection is not as straightforward as it appears.

Furthermore, the recent discipline of agroecology calls into question the idea of agribusiness as efficient or productive. Agroecology attempts to encompass the scientific discipline of researching food production within an ecologically aware framework, together with a politically participative approach that honors indigenous knowledges and tackles practical, real-world problems for the benefit of people rather than multinational corporations (Levidow et al. 2014; Shiva 2016). The landscapes that an agroecological approach produces are very diverse, dependent on context, but would be closer to the values and aesthetic qualities of the human-nature co-creation of the more traditional farmstead.

It may be possible to find positive aesthetic qualities in industrial agricultural landscapes; however, appreciation remains very difficult if we know what lies behind the qualities we perceive. Although we agree with Carlson that it may be possible to find the new agricultural landscapes aesthetically appealing, he underestimates the moral pressure placed on aesthetic appreciation, where that pressure may make it difficult to approve of a type of activity that we know to be harmful. In the case of industrial agriculture, information about the harm caused makes appreciation difficult in the sense that we may feel we have to turn away or we may feel in some sense guilty for finding the colors and forms appealing. In other words, the value of the aesthetic qualities does not itself change, but moral considerations may discourage us from condoning the actions that underlie the qualities perceived.[2] Given these concerns, not only is it quite challenging, but it would not be environmentally beneficial were we to try to develop an aesthetic taste for the new landscapes.

Our objections to Carlson do not, in principle, challenge the view that in aesthetic appreciation we are often able to adapt to changes in the look of the land, and we may have very good reasons for doing so, just as that is required if we are to accept the challenges of new art forms. It is often the case that encouraging more ecologically sound environments means having to adjust to great changes in the aesthetic qualities we enjoy. While sheep farming contributes to the distinctive pastoral quality of the English Lake District and supports a certain way of life, many conservationists will say that it is a barren landscape, with low biodiversity value. Recent "rewilding" projects and their proponents across the United Kingdom agree (Monbiot 2013). Others, including visitors to the national park there, would prefer the sheep to stay, in so far as moving away from this type of farming may lead to changes in the look of the landscapes they cherish. This conflict is complicated by the fact that, as an agricultural practice, small-scale farming (including hill farming) is generally much friendlier to the environment (even though from some perspectives it will be viewed as not as efficient as industrial farming). For example, existing woodlands are preserved, and hedges between small fields provide important habitats for birds and other animals.

As industrial farming has evolved, direct involvement with the land has been replaced by technology, alienating humans from nature. Such change has been viewed as positive by many agriculturalists, for if one does not consider environmental harm, the new agricultural landscapes are much more productive, producing cheaper food more easily (Worster 1999). It is interesting to compare this type of change to picturesque theory and design that, in its more hubristic iterations, has been viewed as "improving" on nature. Likewise, industrial farming could be seen as an improvement in methods of farming, but the result is a landscape out of touch with both humans and nature.

In her study of our aesthetic taste for the pastoral, Sally Schauman (1998) suggests a useful analogy between farming and gardening. Both practices involve cultivation of the land, and a potential result of this activity is a caring relationship toward the environment. One of the reasons we enjoy gardening is because it engages us in a caring and nurturing relationship, in this case with plants rather than daughters and sons or brothers and sisters. Of course, not all gardening and agriculture engage us in this way. Humans have set out to master the forces of nature and bring them under control throughout history—industrial farming is a current example of this quest, and indeed, some forms of traditional farming may also be cited here. But Schauman wants to point out, and we think rightly, that gardening and other forms of cultivation fulfill a desire to participate with natural forces, to experience how the environment reacts to our involvement with it, and to feel satisfaction when nature thrives in our care. Rather than an attempt to control nature, cultivation is a "quest to connect directly with the earth" (Schauman 1998, 184).

Although agriculture is on a larger scale than gardening, this direct connection to the land may be established through some types of farming, such as organic farming, where farmers set out to work symbiotically with natural processes and seek to produce crops that are less harmful to both the environment and consumers. There are some traditional agricultural practices too, which support this type of nurturing and participatory relationship, and we turn to them in the next section.

Schauman's garden analogy suggests benefits for both humans and the environment. We are familiar with the therapeutic effects of spending time in natural areas (an important motivation behind conservation), but gardens, a type of cultural landscape, also present a therapeutic environment through an activity in which self and nature intermingle. "The processes of digging in the soil, smelling the earth, touching plant materials, and watching plants grow seem to have calming and positive effects on most. Gardeners know gardening is a form of meditation. Recently, environmental psychologists have proposed a restorative theory to explain our need for the natural landscape, the nonbuilt environment" (Schauman 1998, 185).

Very close contact to the earth like this is not typical of many forms of farming, but we want to suggest that it may be held up as an ideal that agriculture may exhibit a harmonious dialectical relationship if it follows the lead of gardening.

Practicing agriculture like gardening, that is, through a nurturing relationship to the land, will be good for the environment, but it is also likely to have beneficial aesthetic effects—the cultural landscapes generated will be more on a human scale, more to our taste for the gentle countryside and probably more interesting in the variety that is created. As cultural landscapes, Schauman highlights how gardens are comfortable, "reassuring" environments on a human scale, "places where humans fit in easily" (Schauman 1998, 185–86). There is a wealth of interesting recent literature showing how harmonious human-plant relationships may develop through horticultural activities. In many cases, the relationships that emerge bring with them therapeutic value to humans, which points to the restorative value of nature that comes through certain activities of cultivating the land (Parr 2007). Agricultural landscapes are always on a larger scale than gardening, but they can also be more pleasing when they express a sense of place that is more familiar and less alien. This is perhaps best encapsulated in the agroecological approach or something like the biodynamic farming idea of the farm as an organism, with the geology, weather, plant, animal, and human interactions in balance and supportive of each other and expressing the unique context of the place in which it has grown (Petherick 2010). In our discussion of gardens in chapters 2 and 3, we found grounds for relationships, certainly, and these ideas can be useful if we think of agriculture as gardening on a large scale. This "person-plant-place relationship" can develop into a caring attitude toward place and grounds place attachment; as one of this book's authors has written, "The most powerful way of establishing co-nurturing relationships is by engaging with plants first-hand: planting seeds, nurturing growth, learning about their needs and shaping their and our environment through such interaction" (Brook 2003, 232).

TRADITIONAL AGRICULTURE AND AESTHETIC VALUE

Although not as efficient in a purely economic sense, some kinds of traditional and small-scale farming are not only sustainable, environmentally desirable in terms of encouraging biodiversity, and potentially working in harmony with local ecology, but they also have other beneficial effects, including encouraging a strong sense of community (Jackson 1992; Worster 1999). The positive aesthetic value of traditional farmlands is evident not only in their pastoral beauty, variety, their familiar appeal in terms of the kinds of values they represent, but also in the harmonious dialectical relationship they represent. To illustrate this, we would like to consider three practices that belong to traditional agriculture in Britain: hedge-laying, coppicing, and dry stone-walling. These practices have a practical aim, and they also have an aesthetic component, both in the activity of production and in the appreciation of the combined "effort" of humans and nature. The harmonious relationship may be identified in all of these contexts: (1) in the expression of human aims in the shaping of natural materials, and in the ways that natural elements respond to their manipulation through new types of growth; and (2) in the aesthetic response to the practical result—a woven hedge, coppice woodland, or a stone wall. These practices generate features that are fitting with the look of the land, and they are situated within an aesthetic taste for the countryside. The so-called natural landscapes of Britain and other parts of Europe are, in fact, largely modified by humans. Recognizing that this is the case is a first step to understanding the interplay of humans and nature within these landscapes, and the aesthetic bases and value that emerge through that interplay.

Hedge-Laying

Hedgerows and stone walls are a familiar sight across Britain and in some other European countries too (see figure 4.2). They are pleasing to the eye for the patchwork patterns they create across rolling hills and mountainsides and, up close, for the distinctive textures of woven branches or richly patterned stone. Their primary purpose is to enclose and divide the landscape into fields for various purposes, such as raising sheep, but they have aesthetic qualities as well. In the activity of creating them, we can also discover a harmonious relationship between humans and nature.

Hedge-laying, or hedging, the activity of creating hedges and hedgerows, also involves manipulating plants—such as hawthorn, *Crataegus monogyna*; blackthorn, *Prunus spinosa*; holly, *Ilex aquifolium*; beech, *Fagus sylvatica*—to serve the specific function of a wall or enclosure (and sometimes for the production of wood, as in coppiced hedges). These walls, however, are living. They are created by a hedge-layer, who pleats or weaves together bushes or trees at an angle as they grow in order to encourage denser and connected growth. As Paul Cooper describes the practice, "The shoots or trunks of young trees are partially severed, bent over almost horizontally, and woven through vertical wooden stakes inserted into the center of the hedge. In

Figure 4.2. Hawthorn Hedge, East Lothian, Scotland.

Britain, the dialect word "fletching" is often used to describe this technique" (2001, 42, 45).

Because they form a thick, living structure, hedgerows are an ideal home for a rich variety of species, sometimes rare, including butterflies and other insects, birds, badgers, and other small mammals such as voles, shrews, and dormice. In open fields where there is little woodland or shelter, hedgerows are an important habitat, and in addition, they serve as windbreaks to promote earlier crop growth and help to control soil erosion. Besides their wildlife value and landscape/aesthetic value, hedgerows have historic value, for many of them have been continuously managed at least since the eighteenth century. Richard Mabey claims that the majority of hedges in Britain actually originated before the Enclosure Acts of the late eighteenth and nineteenth centuries, going back as far as the Bronze Age (Rackham 1977; Mabey 1980, 133). Because intensive farming and development have led to the destruction of hedges across Britain, there are now special incentives for farmers to conserve and restore them.

Hedgerows have an interesting history as a particular kind of meeting between culture and nature. Although many hedgerows have been planted, some were entirely natural and, according to Mabey, originated as "strips of natural woodland which were left after a field had been cut out of a forest. Others were the result of shrubs naturally colonizing the lines of staked, 'dead' hedges that were features of the landscape even before the Normans arrived" (2000, 123). "In regions where trees were scarce, as on the poor soils of Cornwall and west Wales, it was often a wall of gorse-covered turves flanked by the ditch from which these had been dug" (Mabey 1980, 122). Hedgerows are the result not only of humans working with plants to create enclosures through the practice of hedge-laying, but also of taking advantage of existing natural structures and maintaining them to meet particular agricultural needs.

Hedge-laying requires special skills. Knowing how to respond to the plant's growth and pleat together branches requires, again, an intimate understanding of these living materials and their environment. The activity involves knowing how to create a living wall without gaps so that, for example, sheep may be kept from wandering. Here, too, there is no explicit aesthetic aim, only a functional one, but in the activity of hedge-laying and the hedgerow itself, aesthetic value may be found. The hedge-layer may be required to think creatively, to consider how a certain pattern of weaving will bring about a livestock-proof hedge. Woven, fresh wood creates a delightful texture through the variety of branch shapes and the density of leaves. There is endless variety in the shapes of branches as they weave through each other, at different angles and with more or less density in the appearance of the structure. The different textures, smells, and colors of the flowers and berries of the various plants create interest. Hedgerows are also active environments: the fluttering of insects and birds, and birdsong, such as a robin's chirp, add another layer to the aesthetic impact. Much of the time one cannot see the birds or insects very closely, as their movement is nestled and hidden within the hedge itself—it's just that the hedge seems alive with activity.

The hedgerow environment that is shaped in the meeting of intentionality and organic processes is quite distinctive, a microenvironment that concentrates the aesthetic appeal into an area that can be experienced up close and intimately. There is also a sense of fascination that all of this interest exists within a shaped, living thing. This is nature within a human modification, and this nature-artifact synthesis is responsible for much of the felt aesthetic satisfaction. Natural processes resist human modification: growth is the constant which the hedge-layer continuously faces.

There is a distinctiveness and particularity in each individual hedgerow environment, as well as in various regional types. The differences between individual hedgerows and different types will affect their aesthetic character because they will have varying sets of aesthetic qualities. Factors determining this character include age, purpose and setting, height and density, mixture of plants, and variety of species living in the hedgerow environment. Mabey points to the possibilities in regional variety:

> The beech hedges round the windswept edges of Exmoor, and the even taller ones rooted like mangroves along the banked lanes of Somerset's Blackdown Hills; the hedgerow hollies planted—or simply tolerated—as ploughing guide-posts in East Anglia; the tall, double shelter belts of the Sussex Weald known as "shaws"; the hedges that are all that is left of the whole woods, and which are evocatively known as "woodland ghosts." (2000, 123)

A purely artistic tradition has evolved out of hedge-laying practice where artists weave together live and dead branches to create wonderful forms. The roots of this art also lie in other practices that work with the branches or other parts of trees, such as basket-weaving and coppicing. These living arts range from environmental artworks to architectural and garden features, such as Mick Petts's "Willow arch" (Wildfowl and Wetlands Trust Centre, London), a Gothic-style arch made from living willow branches woven together (Cooper 2001).[3]

Aesthetic considerations are not a primary concern of the hedge-layer (or waller, as shown in the final section), but this is not to say that aesthetic sensitivity may not form part of the situated experience of agricultural and other work. Assuming the rural worker has little interest in aesthetics leaves one open to accusations of a bourgeois aesthetic (as if rural dwellers were incapable of aesthetic sensitivity), while assuming that aesthetic interests are strong might underplay the hard work and key utilitarian concerns of agriculture. In an attempt to articulate an "aesthetics of work," Justin Winkler uses the concept of "professional sensitivity" to pin down the role of aesthetic experience within rural work in particular:

> *Professional sensitivity* is about a capacity of sensory experience that is present in everyday perception, as well as in professional contexts of performance. This is especially true in the rural context where lived space and worked space are highly congruent, and technical skills and sensitivity amalgamate. Skill is the capacity to repeat a specialised practice on the basis of past experience, diagnosis of a development, and anticipation of a desired effect. Underlying such skills is my assumption that aesthetic appreciation is borne from a search that is an expression of experience, discrimination, *and* practice. (2005; emphasis in original)

Working closely with the land involves a multisensuous awareness and leads to an acute aesthetic sensitivity for a particular environment. Professional sensitivity helps to locate the role of aesthetics more generally in the context of the agricultural practices we discuss here, but it also describes a dimension of the shaping of dialectical relationships between humans and nature.

Coppicing

Our second human-nature landscape and traditional practice is coppicing. The coppice system takes advantage of the regenerative properties of trees. If young trees are cut close to the ground, several shoots will grow up from the severed trunk (called a stool); if these are cut close to the ground, the stool will send up even more shoots. This natural response of the tree to human intervention in its growth was traditionally used to produce long straight poles of wood that could be used for many purposes as well as easily gathered firewood and animal fodder. The number of years between harvesting the poles depends on the species and the intended purpose but is usually at least five and no more than twenty years. The tree can remain productive for centuries and indeed there is evidence of coppiced lime *Tilia cordata* several thousand years old (Mabey 1996, 118).

The virtue of coppicing is the fast production of useful sizes of wood (poles) for making trackways, buildings, fences, and tools and for providing firewood and harvests of leaves for animal feed. The species used are those naturally occurring such as oak, *Quercus petraea*; lime, *Tilia cordata*; hazel, *Corylus avellana*; ash, *Fraxinus excelsior*; maple, *Acer campestre*; alder, *Alnus glutinosa*; and willow, *Salix caprea*. Interspersed among the coppice stools are often individual trees (standards) left to grow to their full height.

Although coppicing is widely practiced around the world, we will take traditional coppicing practice in England to explore the human-nature dialectic of a practice that has been responsible for the structural elements in traditional agriculture (see figure 4.3). An area of woodland is divided into sections and cut in rotation. However, the system is very flexible allowing individual stools to be part harvested or left longer for specific building projects or other anticipated needs of the farmer or woodsperson. The soil type and climate conditions of individual woods and changing techniques over time have meant that every wood is different and, historically speaking, only where paper records are extant is it possible to see a particular regime exactly (Peterken 1993).

The evidence from old stools and pollen records suggests that little planting was done in the practice in the past. Suckering (where trees send up new shoots near base of the trunk) and layering (where a low hanging or damaged branch roots itself in the ground to produce a new tree) would have occurred naturally. If the person managing the wood wanted more stock, he or she needed only to bend a shoot over and fix it under a covering of soil; this was called "plashing." This technique of plashing emphasizes both human-plant entanglement and the mutability of the different parts of the plant. Any leaf node finding itself buried in soil will proceed to produce roots in response to the new environment.

Figure 4.3. Coppiced Trees, Newbridge Wood, England.

Management regimes in the past could be very basic or could involve allowing domestic animals to graze under the mature coppice areas and pigs to forage for acorns or beech mast. It was important to keep stock out of areas in the first few years of new growth as this would be grazed. The most common way of preventing animals grazing the new shoots of coppiced woods was to dig a trench and deposit the earth on the woodland side of the trench to make a bank. This would prevent domestic animals and deer from entering the woodland area. Evidence of huge earth workings around ancient woodlands shows how crucial this aspect of management was. Today, evidence of the eroded banks remains as semi-filled ditches around woodlands and as serpentine lines of smoothed humps and dips in open fields.

The management practices of the past have shaped the type of woodland that exists today, particularly the distribution of plants. Although these practices seem much less intrusive than, for example, plantations of single-species trees and clear cutting, they did leave a legacy. This can be seen easily with the wood banks but coppicing also has an impact on the micro-climate of each area of woodland. The changing patterns of light density in a traditionally managed coppice meant that a wide variety of plants could grow there. The amount of light can change from 5 percent, prior to cutting, to almost 100 percent just after cutting (Peterken 1993, 52). This means that the shade-liking plants usually flower well in the next year and then die back (perennials to a vegetative state) until the shade starts to return as the coppice shoots grow. The sun-liking plants sprout from dormant seeds and flower well for two or three years. The dense shade of the last two or three years of each cycle prevents invasive grasses and shrubs from taking over (Lane and Tait 1990, 77).

Although there is little traditionally managed woodland now, woodland with a long history of coppicing tends to have very rich flora and fauna and is, in some cases, the only place in England where some species survive. This management system is generally recognized as increasing the biodiversity of both flora and fauna. What are the aesthetic qualities of these types of woodland? The characteristically rich bird song and occasional sightings of animals add to the beauty offered by the plants, but it is to the plants that we want to turn. The variety of species is an often-cited feature, both variety of trees and the flora which the changing light density allows to flourish. Variety in the coppice woodland needs slightly more explanation because of the additional cyclical pattern of cutting back small areas. This means that as well as the seasonal cycle of winter, spring, summer, autumn, and all the changes that these bring to deciduous woodland, there is a seven-, eight-, nine-, or up to a twenty-year cycle of cutting and growth with the related changes in the ground flora and fauna. Moving through the changes of light intensity helps to focus one's attention on the exact quality of light around the different tree species.

The growth of new poles close to the ground means that the visitor to the woodland can experience in close proximity the leaves of tree species that in mature forms are far overhead. The particular beauty of unfurling leaves in spring can be experienced very close at hand. We can see their intricate forms and the subtle changes in color as they mature.

The strong vertical gesture of the coppice poles can create an uplifting effect, particularly when seen in proximity to mature trees with serpentine forms such as the beech or gnarled, twisted oak boughs, or when echoed by clumps of ferns. The juxtaposition of the rugged cracked base of the coppice stool and the smooth gleaming bark of the poles is also striking.

Perhaps the most controversial form in the coppice woodland is the newly cut stool, which to the uninformed eye appears to be the stump of a tree cut down and destroyed. If seeking to find beauty in this type of woodland, one could just ignore that phase and say it is necessary but unpleasant and choose to walk elsewhere. However, an understanding of the processes involved in this form of regeneration can, for the careful observer, transform the newly cut stage into a thing of beauty. The internal growth patterns of rings are revealed and bring a new pale color to the woodland floor. Around the edge are the growth points where in a surprisingly short time new shoots will appear and the low compact form will be transformed. The dense compact stage, with the growth seemingly held back for a moment, when seen as part of the process, can be experienced as the most intense expression of the regenerative possibilities of these deciduous trees and the source of the more accessible beauty to follow.

The intense expression of regeneration among coppice, under the stately spread of occasional standard trees and the burgeoning variety of flora, allows one an insight into the nature of woodland. Each of the plants can be seen as arising out of thickening of vegetative space and each forming the cradle of each other's becoming. The knowledge required for this transformation of appreciation to take place is not necessarily scientific knowledge in the frameworks of either biology or ecology. It can be assisted by the craft knowledge of woodland management but can spring from attentive observation with no external epistemological basis beyond the trees themselves.

The beauty of coppiced woodland usually features as an afterthought to claims of its ecological benefits, heritage value, or potential usefulness. There is also, we would maintain, a beauty in the interweaving of human utility and the seeming generosity of trees, where the person engaged in coppicing is both finding the exact "product" they want for this specific occasion and aiding the future flourishing of the woodland and future humans.

Stone-Walling

In their basic materials dry stone walls are not living things, but they do acquire organic components through time—lichen, ferns, and insects living among the stones. Here, too, the waller has no explicit aesthetic aim but sets out to build functional walls that can withstand the elements and sheep's attempts to climb over them. Presentation is an implicit concern, but walling is such hard work that the concerns of beauty are not close at hand.

However, there is an unintended beauty in the patterns that result in the walls themselves from the need to place stones in such a way as to achieve a good fit to produce a solid, strong form. The patterns of the stones are not planned except in so far as each stone picked up must fit well into the place presented by the previous stone. In this respect, the beauty of the walls may be described in terms of "functional beauty," where both the aesthetic qualities and function of the walls emerge jointly with one another.[4] John Benson explains how these aesthetic qualities emerge:

> First, the way in which the irregular surfaces of individual stones are justified, like the lines of print on a page, so that one gets a diversified surface which has a simple and uniform contour, especially when viewed obliquely from a little distance. Second, the complex pattern and variegated texture that results from the close fitting of stones of different and irregular shapes and sizes. (1999, 4)

Irregular stones and surfaces on the stones themselves are brought together into a stunning result: usually, a clearly patterned and perfectly shaped wall, an effect achieved through arduous, heavy work in selecting and placing each stone. As Dan Snow, a waller, has put it, "When I start building, I don't set out to make a beautiful wall. My intention is a lasting structure. Throughout the day, I apply my efforts to the basic principles of drystone construction. If I follow them successfully, I'm likely to leave behind a pretty good-looking wall" (2001, 29).

The texture of stone walls varies according to the type of stone and region. For example, in the Lake District, slate is commonly used instead of the limestone of the nearby Yorkshire Dales. The effect of slate is a cleaner, smoother line and surface, and a gray-green color, compared to the rougher, more rugged appearance of the Dales walls. In both cases, stones are typically collected from a local quarry, stream beds, or fields nearby where the wall is built (particularly if they need to be cleared from the fields), which means that the walls themselves blend naturally into the landscape. As with hedges, some stone walls evolve directly from natural features of their situation, although not literally in an organic sense (see figure 4.4).

Figure 4.4. Stone walls and barn showing use of stone from surrounding site. Yorkshire Dales, England.

The same type of stone used for walls is also, in many cases, used for barns and houses in the area. The textures of the walls are aesthetically pleasing close up to both eye and hand, but also from far away, where the overall patterns may be viewed more easily. From further away, the configurations of several walls emerge, as they traverse, snake, and intersect across vast expanses of green dotted with white sheep and sometimes, extraordinarily, disappearing over a mountainside.

The distinctive aesthetic appeal of these walls, such as their tightly formed pattern, is underlined when they are compared to other stone walls, for example, those found in the main agricultural area of Norway (in the southwest). These walls have the same function, to enclose fields, yet they seem to serve the equally important function of providing a place to pile up and recycle the numerous boulders that litter the fields. The stones used in these walls are much larger in size, and they tend to be more spherical, which gives them a much rougher, unfinished appearance (indeed, the stones are so large that they have probably been moved with the help of machinery or horses). They cannot be arranged closely together but rather sit on top of one another in a jumbled sort of way—a length of piled up boulders along each side of a field.

Wallers, like the hedge-layers, must know their materials extremely well. Finding the right stone at the right time is at the center of this practice, for to make mistakes is to waste valuable energy or to create a wall that will collapse (Benson 1999, 3). There is a dynamic relationship here too, as the waller responds to the shape of each individual stone placed into the wall and chooses another, specific shape to fit into the form of the wall as it progresses from one stage to the next. The waller is not responding to or manipulating organic processes, but instead, there is a search for a natural form, from the land, to fit into an intentionally shaped form. While the relationship here is chiefly harmonious, the heavy weight of the stones and liability to collapse present ways in which nature resists human manipulation.

One might characterize the stones' side of the relationship in terms of how each stone's shape meets the specific requirements of the waller through its particular features. Snow captures the live quality of stone: "Stones seem to want to congregate. In a stream they remain together when all else is washed away. On a mountaintop they endure when everything else is blown away. Every piece finds its right relationship to every other" (2001, 91). The relationship between waller and stone may be brought out further through a key design principle of Japanese gardens, *kowan ni ishitagau*, or "following the request." A main stone is placed and, in harmony with the stone's "native characteristics," each successive stone is chosen to complement them. As Yuriko Saito describes it,

> According to *Sakutei Ki*, the gardener "should first install one main stone, and then place other stones, in necessary numbers, in such a way as to satisfy the

> requesting mood of the main stone". The unity of the whole arrangement is achieved by the selection and placement which accentuate the characteristics of each stone through contrast rather than repetition. For example, the *Sakutei Ki* suggests that the "running away" stones should be accompanied by "chasing after" stones, a leaning stone by a supporting a stone, a "trampling" stone by a "trampled stone", a "looking-up" stone by a "looking-down" stone, and an upright stone by a reclining stone. (1996, 49)

As with the woven trunks and branches, dry stone walls have been constructed for more artistic aims, such as a feature in gardens, or as actual artworks themselves. Andy Goldsworthy a notable here as he has created several environmental artworks fashioned after the stone walls of the UK countryside. His stone works range from cone-like shapes to sheep folds, winding walls through forests, and walls disappearing into the sea (Goldsworthy 1994, 2000). David Matless and George Revill describe how Goldsworthy conceives of his stone artworks as "monuments to walling," where "such works sustain the processes to which they refer, not to become future memorials to lost vernacular arts" (1995, 433).[5]

THE LEGACY OF HEDGES, COPPICES, AND STONE WALLS

Despite our arguments in favor of traditional forms of agriculture such as stone walls and hedgerows, it is important not to romanticize them. As Schauman argues, even the attractive pastoral farmlands of rural America hide some forms of environmental harm, such as degraded water quality (1998, 188). While hedgerows are considered desirable for several reasons, ecological and otherwise, their history is not altogether innocent. Many hedgerows and stonewalls are the legacy of the Enclosure Acts, when wealthy landowners enclosed common fields, creating private property that cleared the land for sheep farming. This badly affected rural communities, resulting in the loss of land, livelihood, and other deeply adverse changes to their way of life (Mabey 1980; Cosgrove 1998; Herrington 2008). In response to these concerns, it would be a mistake to associate stone-walling and hedge-laying too narrowly with the Enclosure Acts, since these practices existed prior to the eighteenth century, and they continue today within a new narrative. That new narrative is not necessarily characterized by farming managed by a few wealthy landowners. Rather, it is now set within a wider context that includes many positive values associated with small-scale farming, organic farming, and the benefits to the environment that follow. However, again, we do not want to overlook the problems that are well known within this new narrative. For example, the foot and mouth epidemic in Britain in 2001 refocused attention on the hardships and economic fragility of hill farming in areas such as Cumbria and Northumberland. Alongside this, in recent discussions about rewilding in the United Kingdom, criticisms have been made of the biodiversity deserts created by sheep farming across the country (Monbiot 2013; Brook 2018).

In the case of coppicing, it could be argued that instead of coppiced woodland there could be more natural woodland or pockets of wilder areas. Also, most of what was coppice woodland in the past is now overgrown and would be classed as lapsed managed woodland. Many of the traditional uses of the poles have gone as imported animal feed and "cheaper" clear felled timber and charcoal are shipped around the world. Despite the biodiversity benefits of traditional coppice with standards, the main resurgence of coppicing is in short rotation coppice to produce fuel for power stations. However, this is a single species, cut every four years where undergrowth is kept down with weed killers, thus losing many of the ecological benefits and, as a monoculture, it is not seen as having much aesthetic value (Sadler 1993).

Our views on traditional agriculture are also open to the criticism that the practices discussed are relatively minor compared to other agricultural activities. In response, we would argue that hedge-laying, coppicing, and stonewalling reflect an ideal, like the ideal that gardening presents for developing a direct connection to the land. By turning back to these practices, we may be reminded of the harmonious relationships and fully integrated local systems that still exist in some parts of the countryside, and they illustrate good examples of the particular ways in which these relationships may develop and express themselves.

In conclusion, we would like to point to another aspect of the human-nature relationship that ought to be carefully considered. It is worth emphasizing that the more harmonious relationships are not necessarily those in which the product of human-nature interaction appears more natural or when there is less human involvement. Agriculture and other types of cultural landscapes involve practices where humans shape and modify nature in explicit ways, and this can be a valuable enterprise rather than something we want to discourage as an intrusion upon nature. Such relationships, while not always harmonious and sometimes harmful, enable interactions with environments that potentially deepen attachment and increase understanding of both ourselves and the natural world. Rather than romanticizing the human-nature relationship, we can use it in a few fruitful ways: to help in understanding aesthetic appreciation of agricultural landscapes and to sort out the harmonious from the deeply conflicting relationships, to discover how positive aesthetic value emerges through harmonious relationships, and finally, to help in thinking beyond the aesthetic context to reflect on human-nature relations in the broader context of cultural landscapes and other kinds of relations besides the aesthetic, such as the ethical.

NOTES

1. There are other forms of farming that are potentially more ecologically friendly, such as natural farming, permaculture, biodynamic farming, and so forth. This will also hold for forms of agriculture which involve animals, for example, some free-range and organic chicken farms, some free-range pig farms, natural versus intensive and conventional beekeeping, and so forth.
2. We should emphasize that moral considerations are likely to make appreciation difficult in some ways, but this does not affect the merit of the actual aesthetic qualities found in the landscape. Elsewhere, we have argued that aesthetic value is autonomous from moral value in so far as moral defects do not count as aesthetic defects (Brady 2003). The relationship between aesthetic and moral values is complex, and we cannot discuss this in further detail here; for a useful discussion, see Eaton (2001). In chapter 7 of this book, we discuss other issues with respect to the relationship between aesthetics and ethics.
3. See chapter 6 for further discussion of Petts's work.
4. We develop the notion of "functional beauty" here differently than Parsons and Carlson (2008). For further discussion, see chapters 1 and 6.
5. For a discussion of different practices using stone as the medium, beyond cases in the United Kingdom, see Patrizio (2011).

Chapter 5

Environmental, Land, and Ecological Art

In this chapter, we explore practices of environmental, ecological, and land art-making and how they fit into the pragmatic, relational turn in environmental philosophy. What kinds of relationships emerge between humans and nature through artworks that synthesize human intention and actions, and in some cases technology, with natural processes and materials? We approach the problem from the artistic rather than appreciative side and focus on a set of artists who have been given attention within recent discussions in environmental aesthetics. We begin by exploring how these kinds of artworks are understood, before moving on to a particular type of aesthetic objection raised against them. While we recognize that some of these artworks constitute an "aesthetic affront" to nature, others potentially involve a more positive relationship between humans and nature. To give content to that positive relationship, we shall argue that some artworks express "aesthetic regard" for nature.

UNDERSTANDING ENVIRONMENTAL, LAND, AND ECOLOGICAL ARTWORKS

Environmental and land art ranges from small sculptural objects to grand gestures in the land and from impermanent ephemeral works to permanent ones (Ross 1998; Andrews 1999). Some works appear indoors in art galleries and museums, but these forms of art are best known for being site-determined and in the land or environment itself, that is, outdoors within more natural settings. Although sculptors and other artists use natural materials all the time, environmental and land artists will often use nature as material, subject, and setting.

Environmental and land art emerged in the early 1960s, and its genealogy can be traced back to a number of artistic movements and art forms from the twentieth century, including minimalism, postminimalism, public art, conceptual art, process art, interventions, happenings, the "Arte Povera" movement, and installation art. There are also roots in gardening, landscape design, and other human modifications of the environment. Robert Smithson pointed to the affinity between his earthworks and the earthworks of Central Park in New York (designed by Olmstead and Vaux), and much earlier, to the earthworks of picturesque gardens in the eighteenth and nineteenth centuries and ancient artistic and ritualistic earthworks such as land drawings, megaliths, and mounds (Andrews 1999). Although it is difficult to stipulate the true beginnings of the genre of environmental and land art, some point to an important convergence in 1968, marked by the first solo shows of Richard Long and Giuseppe Penone in Europe and the exhibition, *Earthworks*, at New York's Dwan Gallery.

In art history, Malcolm Andrews (1999, 205) has grouped various artists and works into different categories. "Earth art" and "land art" include minimal and ephemeral interventions in a site (e.g., work by Richard Long, Andy Goldsworthy, and Michael Singer), large-scale sculptural earthworks on site (e.g., Michael Heizer and Robert Smithson), "unmediated installation in an art gallery of materials gathered from a landscape site" (e.g., Walter De Maria and early Smithson), "landscaped reclamation or planned naturalization of industrial wasteland" (Smithson), and "acts of conservation of natural land that involve decisions about what traditional usages of the land are retained" (e.g., Alan Sonfist and Agnes Denes). Stephanie Ross, a philosopher, formulates these categories:

(1) masculine gestures in the environment: Heizer, Smithson, De Maria, Turrell
(2) ephemeral gestures in the environment: Singer, Long;
(3) environmental performance art: Boyle, Fulton, Hutchinson, Christo
(4) architectural installations: Holt, Aycock, Miss
(5) didactic art: the Harrisons
(6) proto-gardens: Sonfist, Irwin, Finlay
(7) sculpture gardens and parks

(1993, 170; see also Rosenthal 1983).

In our analysis below, we draw liberally on these two approaches, though we add a layer of works that can be described as more explicitly "ecological."

Is it possible to define this group of art forms? Given the general approach of this book, we cite here a few definitions from work in environmental aesthetics. Carlson defines "environmental art" as "in or on the land in such a way that a part of nature constitutes a part of the relevant work . . . not only is the site of an environmental work an environmental site, but the site itself is an aspect of the work" (2000, 150). John Fisher adds that the "site should be more or less natural and not an artifactual or urban site" (2007, 280). This definition is too narrow, however, because as Sue Spaid (2002, 10) points out, many land works will be situated in an urban setting, such as Agnes Denes' *Wheatfield* (New York, 1982) or Patricia Johanson's *Fair Park Lagoon* (Dallas, 1981).

These categorizations and definitions point to the significance of the site, or being site-determined, as well as the role of natural materials, so to that extent they all recognize a key similarity. But some works of art may not be exhibited "in the land," as Andrews mentions. Here we find not only installations by the likes of Smithson, but also forms of ecological art, for example, the Harrisons' *Portable Fish Farm* (shown in the Hayward Gallery 1971). Ecological artworks share an interest in ecological processes with some environmental and land artworks, yet they are more consistently biocentric in their ethical underpinnings.

There is some continuity between early land art and later ecological trends. The earthworks artists of the 1960s and early 1970s were, arguably, more interested in taking art into the land rather than the growing environmentalism of their time. However, we do find the Harrisons' work and, for example, *Bog Action* (1971), where Joseph Beuys waded into a marsh until only the top of his hat was showing, seeking to bring attention to the destruction of wetlands in the Netherlands (Beardsley 1984). Spaid categorizes some kinds of ecological art as "eco-vention": "an artist-initiated project that employs an inventive strategy to physically transform local ecology" (2002, 1). Johanson is notable here, but the category potentially includes more recent artists and projects, such as Lynne Hull's trans-species art, AMD&ART's community-driven projects, Brandon Ballengée's species reclamation art, and Fritz Haeg's "Animal Estates," which works with animal clients to create urban dwellings for them.

In response to the great diversity of artworks associated with "environmental," "land," and "ecological" art (also, "earth art," "earthworks"), and the different attempts to classify or define them, here we shall adopt a "family resemblance" approach rather than to seek a definition in terms of necessary and sufficient conditions. We note various similarities and overlaps between different art practices with respect to how they are sited, their materials, their conceptual underpinnings, and ethical motivations, especially where these have a bearing on the character of artist-nature relationships.

Moral and aesthetic-based objections have been raised against some forms of environmental and land art by artists, art critics, art theorists, and, more recently, philosophers, since at least the early 1970s (Gussow 1972; Auping

1983; Crawford 1983; Humphrey 1985; Carlson 2000; Tufnell 2006). In the next section, we revisit from chapter 3 a particular type of aesthetic criticism discussed by Crawford (1983) and put forward by Carlson (2000), the "aesthetic affront to nature," and consider how it applies to environmental and land art. We shall also discuss some ecological works, why they are important, and how they fare in light of Carlson's criticism.

AESTHETIC AFFRONTS TO NATURE THROUGH ENVIRONMENTAL AND LAND ART

Recall that an "aesthetic affront" identifies human actions that cause some kind of aesthetic wrongdoing toward nature, as distinct from an offense falling more clearly within the moral domain. While it is clear that humans commit offenses against nature in terms of environmental harm all the time, it is useful to distinguish inappropriate treatment that occurs through forms of aesthetic engagement (Godlovitch 1998b). For example, Crawford explains how the land art of Christo and Jeanne-Claude has been criticized as "artifacts [which] forcibly assert their artifactuality over against nature, by their size, their engineering complexity and their synthetic components" (1983, 57). They constitute an aesthetic failure in so far as they are "destructive of their natural setting within the aesthetic context" (1983, 57).

We have seen that Carlson defines the concept of an aesthetic affront as aesthetic rather than moral because it is "generated by the aesthetic qualities of an object, rather than by, for example, its social, moral, ecological, or other such qualities" (2000, 151). Importantly, an aesthetic affront is not just related to the appearance of the artwork, but as Carlson explains, "It is a function of changing an object's kind and thereby altering its aesthetic qualities" (2000, 155). This type of affront is not synonymous with environmental impact; rather, it is more like an "aesthetic indignity" or an "insult" against nature, where nature is affronted even if there is no consciousness of an affront being made. Carlson argues that several artists are open to this charge:

> Perhaps "defile" is too strong a word to characterize most environmental art. Nonetheless, the general way in which environmental artists alter nature's aesthetic qualities by turning nature into art does seem to support its being an affront to nature. This is illustrated by Heizer's works such as *Displaced-Replaced Mass* (1969) in which a 52 ton granite boulder is "messed with" by placing it in an excavated depression. It is also evident in works such as Christo's *Surrounded Islands* . . . and *Valley Curtain* (1971–2), 200,000 square feet of bright orange nylon polymide spanning a Colorado valley, or Oppenheim's Branded Hillside (1969), a "branding" of the land executed by killing the vegetation with hot tar. In such cases nature is "redefined in terms of art" at a cost

> to its aesthetic qualities such that to speak of an affront, if not a "denigration," is quite appropriate. (2000, 155)

Carlson specifically discusses what appear to be more extreme cases of aesthetic affronts, but many forms of environmental and land art are included, and he suggests that *most* environmental art is somehow an aesthetic imposition on nature. Later in his discussion, he identifies a few cases that constitute less serious affronts, the most convincing being works by Michael Singer (e.g., *Lily Pond Ritual Series*, *First Gate Ritual Series* 1976) and Alan Sonfist (e.g., *Time Landscape* 1965–2007). These cases, and others like them, are less problematic because they attempt to highlight and bring out nature's qualities, rather than altering them (Carlson 2000, 160).

On a general level, our attempt to defend ecological and other artworks from Carlson's position is shared by philosophers such as Heyd (2002), Lintott (2007), and Boaz Simus (2008). We mention Heyd's position below. For Lintott, we need to consider whether or not environmental artworks are *artistically* "worth it." That is, we should determine, on a case-by-case basis, whether such projects are worth it given a set of considerations, including, ecological harm, the artistic materials and process of production, and the meanings expressed by the work. It may be that some works encourage shared feeling and awareness of environmental harm and the need to protect the environment for itself, for ourselves, and for future generations. Boaz Simus argues that we need to be keenly aware of ecological harm caused by these artworks, and he applies the notion of "ecological citizenship" where environmental art can provide opportunities to develop ecological good rather than ecological harm. Our own position specifically emphasizes respectful culture-nature interactions by focusing on harmonious relationships and adopting the idea of "aesthetic regard" to flesh this out in the context of artistic practice.

To determine whether Carlson's position is warranted, we first address the more problematic end of the scale—the earthworks of Michael Heizer and Robert Smithson. We then examine less problematic cases and nonaffronts by a range of other artists. Although we recognize that an aesthetic affront occurs in some cases, we will argue that if the distinctive types of relationship that develop through artistic engagement with natural forms and processes are explained, many artworks may be said to constitute a form of aesthetic regard for nature. That regard emerges through an interaction of cultural and natural forces, rather than an expression of nature alone or human culture alone.

"Regard" has several meanings that make it an especially useful concept for capturing both aesthetic and moral concerns. The first group of meanings points to ways of carefully attending to or giving something detailed consideration. We are not suggesting that aesthetic experience is limited to *looking* at something, but this mode of experience often begins with attentive perception, especially attending to and engaging with various qualities through a *range* of senses (Brady 2003). Building upon the idea of careful attention, another set of meanings can be tied to respect and esteem: to regard something is to care for, to look after, respect, and show deference toward it (Oxford English Dictionary 2009). Brought together, *careful* attention points to how aesthetic regard is integrated into positive relationships between artists and the natural world.

DEFENDING ENVIRONMENTAL AND LAND ART

Earthworks

All artworks involve the actions of artists, but when working with nature and environment, the artist is decentered to a greater or lesser degree depending on the work in question and to what extent the artist chooses to interact with natural processes. One should not assume that any artist is always in control of his or her materials. Indeed, in many forms of art this will be the case, but with respect to natural materials and sites, it is more likely that there will be shaping, resisting, thwarting, and so on, from both sides of the relationship. We suggest that less harmonious types of artistic human-nature relationships are found in artworks that exhibit a kind of aesthetic egoism, where the artist is not decentered.

Ross describes works by Heizer and Smithson as "masculine gestures in the environment" due to their scale, their inaccessibility, and because "traveling to see them requires braving wilderness, rattlesnakes, and the desert's climatic extremes" (1998, 210). Heizer aligned himself with the rugged, pioneering spirit of the American West; like other early earthworks, Suzann Boettger points out that his projects "manifest no connection to the idea of nature or of the earth as a source of living being itself. Instead, the land was viewed as a flat, hard surface on which to boldly make one's mark, even if in panoramic perspective that gesture is dwarfed by the expansive environment" (2002, 114). Some art critics of the time associated the earthworks of Heizer and Smithson "with a kind of macho aggression in which the violation of the earth with huge mechanical diggers was seen as a raw assertion of male authority over Mother Earth" (Andrews 1999, 213).

In an interview from 2015, Heizer remarks, "Years ago, when I had no money and I made a work of art, maybe I couldn't afford to make it more resistant to the weather. I did, however, exploit that situation. I wasn't an environmental, greenie artist making things out of moss and leaves. But I knew that some things dissipate, and I factored that into the work" (Farago 2015). In Heizer's case, the aesthetic affront charge is potentially a strong one. Although site-specific, his works are the clear focal point rather than the site itself. In addition to the example discussed by Carlson (*Displaced/Replaced Mass*), *Double Negative* (1969–1970) cut a linear sculptural form, fifty feet deep and fifteen hundred feet long, into Nevada's Mormon Mesa. This work could be seen as constituting a disruptive relationship with the environment, akin more to a power relationship. Heizer and many other earthworks artists were less interested in the growing environmental movement of their time and more interested in taking art outdoors and outside the confines of the gallery. An alternative reading would see his works as pointing to the imbalance of power between human and nonhuman rather than actually reproducing that imbalance.

In another recent interview, Heizer points to the significance of the land for his massive, in-progress sculptural project, "City," situated in the desert landscape of Garden Valley, Nevada (also his home, after leaving New York City some fifty years ago). He appears more sensitive to the natural context of this work and site-specificity, "Over the years we saw how the thing stood the test of time, what didn't work, what had to be rebuilt, what happened when the valley flooded, in different climates. It's like a handmade object now, erased, redone, adjusted, not just fabricated. It's part of nature, here for the millennia" (Kimmelman 2015). Time spent in Garden Valley has very likely enabled Heizer to cherish the place, rather than to treat it merely as a canvas. Although the sculptural forms of "City" stand out in the landscape as human-made, the site is very important. So, while Heizer's concerns are not overtly environmental, his recent remarks suggest a deep interest in the desert situation of "City" and offer a reply, if somewhat mild, to the affront charge. In the context of discussions of the proposed geological age of the Anthropocene, Heizer's works might be said to present an artistic, relatively small-scale case of humanization of the environment.

Carlson makes his aesthetic affront objection clearly against Heizer, yet it is qualified against Smithson. The objection is concerned with *nature* affronted through the alteration of natural qualities by an artist so that when an environmental artwork does not alter these qualities, as in cases where an artwork draws on and is situated in a more nonnatural site, the affront *to nature* doesn't occur. Thus, he contends that works by Smithson in postindustrial sites or less natural sites will not necessarily constitute an aesthetic affront. But this is not an easy qualification to make given the ontological complexity of the sites in question. A former quarry mixes natural and cultural qualities and processes, and although Carlson is aware of this, it may be difficult to tease out or distinguish alterations that are made to natural versus nonnatural qualities. A second qualification with regard to Carlson's views of Smithson is more straightforward, namely, he recognizes Smithson's more complex relationship with the land.

As both artist and thinker, Smithson was interested in the natural physical processes of entropy and decay, seeing himself as an artist who took on the "persona of a geologic agent where man actually becomes part of that process rather than overcoming it. . . . You just go along with it." (Flam 1996, 298; Boetzkes 2010). His most famous work, *Spiral Jetty* (1970), spirals fifteen hundred feet into the Great Salt Lake near the shore at Rozel Point (figure 5.1). The work is dynamic rather than static: depending on the water level of the lake, it submerges and reemerges, now partially covered with sparkling salt crystals. The artwork operates on various levels, from the sheer physicality of the spiral, to its expansive lake site with long stretches of water and land, to ideas about earth, creation, and metaphysical ideas.

Smithson described the site of *Spiral Jetty* in evocative terms:

> As I looked at the site, it reverberated out to the horizons only to suggest an immobile cyclone while flickering light made the entire landscape appear to quake. A dormant earthquake spread into the fluttering stillness, into a spinning sensation without movement. The site was a rotary that enclosed itself in an immense roundness. From that gyrating space emerged the possibility of Spiral

Figure 5.1. Spiral Jetty (Robert Smithson 1970), Great Salt Lake, Utah.

> Jetty. No ideas, no concepts, no systems, no structures, no abstractions could hold themselves together in the actuality of that evidence. (Flam 1996, 147)

Smithson's more subtle relationship with the earth is worth a closer look, we believe, for it shows how his art can be defended against the aesthetic affront charge in all cases.

Many of Smithson's sites were chosen because they had *already* been damaged by human actions, for example, waste sites and disused quarries (e.g., *Asphalt Rundown*, 1969). As such, his works are in some ways congruous with the humanly altered character of the site, thereby drawing attention to human impacts on nature or more negative relationality. Smithson's art was not in any way attempting to artistically beautify former industrial sites; he argued against reclamation art that covered up the damage done to the environment and aimed, rather, at leaving that damage visible while engaging artistically with the site (Spaid 2002, 54). This suggests integrity in relation to the narrative of that landscape, at least in respect of its more recent past. He did not seek to beautify, nor did he seek to remediate. Furthermore, his projects might be said to have the somewhat ironic effect of inducing concern for nature through art created *within* a destructive environment. Thomas Heyd supports this role for land art:

> Earthworks . . . are like fingers pointing to what little we offer in exchange for nature. They show us this by mimicking the rough handling of wild, natural spaces carried out by our entrepreneurs and whitewashed by our policy makers on a daily basis. The difference between earthworks and industrial interventions in nature is that earthworks, insofar as not otherwise useful, leave the onlooker no choice but to reflect on the place of human intervention in wild nature and, in this way, may lead to renewed attention to the supposedly justified interventions in nature of the everyday. Furthermore, at least in many cases, to look at earthworks is also to look at the wilderness that surrounds them. (2002, 7–8)

Additional evidence for Smithson's more enlightened view of the artist's relationship with environment comes through his many writings and interviews. Smithson later developed a more ecological conscience, urging that "art can become a resource that mediates between the ecologist and the industrialist. Ecology and industry are not one-way streets, rather, they should be crossroads" (quoted in Tufnell 2006, 97). It appears he was moving toward seeing earthworks as having a restorative or remedial role in the environment. Smithson's reflections on his own artistic ideas reveal that his approach, while not environmentalist or nonanthropocentric, is opposed to Heizer's aesthetic egoism in important ways.

In the essay, "A Refutation of Historical Humanism" from 1966 to 1967, Smithson argues that the "artist should resist all forms of humanism" and that "art is not self'" (Flam 1996, 336). He is dismissive of "vain" aesthetics, arguing against self-expression in art (Flam 1996, 334). Of course, one could say that there is inherent vanity in using the natural world for the purposes of art, but this ignores the very sense in which an interactive approach to nature may lead to greater understanding of it, as if not engaging at all would be a better route to valuing environment. Smithson is not a noninterventionist or preservationist, and his views are more akin to how art can interact with geological or other forces. As we have seen, his understanding of nature is deeply informed by an interest in physical forces, geological change, and natural disasters. This point also serves as a reply to Michael Auping's criticism that Smithson was merely interested in "hiring himself out to decorate an area of landscape the mining company had exploited" (1983, 97).

Smithson's approach to landscape was not, clearly, one of aestheticization, especially if we recognize that his interest in both more and less humanly modified sites was positioned within the "unscenic," a historically neglected area of aesthetics and landscape (Saito 1998; Maskit 2007). His "aesthetics of disappointment," a fascination with the unscenic, the ruined and postindustrial, paid tribute to the relentless, inevitable power of nature. But this is not nature held remotely or at a distance; such power affects human actions and built structures. Smithson's *Partially Buried Woodshed* (1970), in which he "took 20 cartloads of earth and piled them on this woodshed until the central beam cracked," was motivated, in part, by an interest in "dealing with the unexpected, and incorporating that into community" (Flam 1996, 305–6). The woodshed resonates with the "buried architecture" of houses and villages buried in ashes from volcanic eruptions, and for some, expresses the ruptures and disintegration felt in late 1960s America. More recently, the posthumously realized work, *Floating Island* (2005), presented a landscaped barge that traveled around New York City's waterways, operating as a floating island of soil, rocks, shrubs, and native trees. In this work, we see, again, Smithson's interest in engaging people with landscape and process, in this case within an urban setting (Yusoff and Gabrys 2006).

Overall, it appears that Smithson's artworks and the ideas behind them are more closely connected with a genuine interest in nature—broadly understood—than might first appear. Do his works express aesthetic regard for nature? Although he was not an ecologically sensitive artist, there is implicit regard at least in the ways his projects connect to and interact with natural processes like growth, decay, and entropy, and in some cases, explicit regard for complex natural qualities as they emerge in works like *Spiral Jetty*. On Carlson's approach, which implicitly favors "wild" nature, Smithson's work will not be embraced. But if we look instead to an environmental ethic such as environmental pragmatism, which recognizes the place of human-nature interactions, it can provide a framework where "hands-on" nature is acceptable, as long as it is benevolent and respectful in spirit.

Performative Art

Christo and Jeanne-Claude's works ("environmental performative art" in Ross's categories) also occur in the land, but the human-nature relationship tends to be oriented toward the human and the artistic. While less "macho" than earthworks, they lean toward emphasis on the artwork and the process of achieving it in the chosen environment. Their use of steel, rip-stop nylon fabric, and other human-made materials to wrap objects and environments gives their installations a strongly artifactual feel in contrast to works that may draw upon a mixture of artifactual and natural materials. Their large-scale installations in the land have been described by Ross as a type of "environmental performance art" (1998, 208). The artists are well known for wrapping islands (*Surrounded Islands* 1980–1983), architecture (*Wrapped Reichstag* 1971–1995), and for using fabric imaginatively through environments (*Running Fence* 1972–1976).

Their projects have also been described as a form of process art because they take years to plan, require planning permission, environmental checks, and community discussions and approval. Taking an example from their recent work-in-progress, "Over the River" has been "postponed" due to environmental opposition from the group, Rags Over the Arkansas River, or ROAR. ROAR argues that the project will have negative impacts on "the headwaters of the Arkansas River, the Bighorn Sheep Canyon, its inhabitants and the communities that depend upon them" (ROAR 2016). Much of the artistic process is community work and diplomacy—including long court proceedings. Once executed, the artworks are usually only installed for a week or two to minimize ecological damage and for other practical reasons.

Although a kind of conflicting relationship is emulated in Christo and Jeanne-Claude's works, it is not one of aggression or power, but perhaps more one of anthropocentrism or "art-centrism." The lengthy process and hype surrounding their projects, and the dramatic ways they transform spaces, means that the artists themselves, and the art, become the center of attention. While they can be seen as showing regard for the sites or environments they choose, the appreciator's attention is ultimately on the forms, colors, and materials of the artworks. The environment serves as more of a setting that inspires their works and acts as a large-scale site for their installations (even if their artworks can be described as site-conditioned and site-specific). In this way, some of their installations may be seen as self-referential in the way Heizer's earthworks are—human creativity imprinted on the landscape. Christo and Jeanne-Claude claim that they want to bring attention to the environmental qualities of the spaces they use or the objects they wrap, and sometimes they do. But the attention remains fully on the work and the process that has brought it to completion. It would be uncharitable, however, to suggest that they do not care about the environment at all, for they seek to minimize ecological damage caused by their projects, and for *The Gates* (2005), a project in New York City, all proceeds from sales of postcards, T-shirts, and so forth were donated to a charity they founded, "Nurture New York's Nature Inc."

Olafur Eliasson's works share some qualities with Christo and Jeanne-Claude in being performative and large scale, but they are more deeply concerned with engaging people with environmental spaces. His site-specific works explore nature, technology, and environment within public spaces using artifactual and natural materials including light, heat, water, moisture, steam, and ice. In *Green River Project* (1998–1999), Eliasson performed "unauthorized" and "anonymous" actions in Stockholm, Tokyo, and Johannesburg, dyeing their rivers green using a nontoxic and nonpolluting colorant. Like the early earthworks artists, he challenges the institutional settings of artworks and, in this work, sets out to encourage urban people to reperceive their everyday environment. In Eliasson's action in Stockholm, the river was spontaneously dyed green in minutes, causing passers-by and people in cars and buses to stop in amazement as the water was transformed to something at once strange and familiar (Rynszteijn 2002).

Ephemeral Art

Ephemeral art, or what Ross (1998) calls "ephemeral gestures in the environment," is exemplified by artists such as Goldsworthy and Singer who, although not environmental activists, adopt a smaller-scale approach with a sensitive use of materials. Their works leave only soft impressions on the land, showing a clear interest in human engagement with the environment; overall, they are more in tune with developing an intimate relationship with nature. In turn, their projects encourage appreciation that is more on nature's terms than the artist's. The artist's role becomes one of enabling or increasing attentiveness to nature's qualities by pointing to them, highlighting them, and working with them creatively. In these ways, ephemeral artworks show aesthetic regard for nature.

In short periods of time, just moments or hours, Goldsworthy creates pieces that interact with the rising or setting of the sun, ocean tides, the wind, or the dynamic properties of water flowing in a stream. Many of his projects are temporary and engage with natural unpredictability: a breeze blows leaves away; the midday sun melts a work crafted from shards of ice. As one of several artists who work mainly with their bodies rather than using tools or technology, he engages intimately with the environments in which he works, using materials from the sites themselves; flower petals are licked together to form brilliant colored shapes or leaves are woven together in a line. Working in conjunction with a site's particular qualities—complexity, simplicity, delicacy, strength, changeability, varying shapes, and textures—he brings out the dynamic possibilities of art and nature through space and time. Natural processes impede and support artistic actions. It is possible to find conflict here, as the stones he so carefully tries to layer into a conical form fall into a heap on the seashore (Riedelsheimer 2004). Nature interacts through resistance, and Goldsworthy must start again with his task. Through this collaboration with natural forces (wind, tides, flowing water), the relationship that grows is one of closeness rather than distance. This is also made possible by the artistic context of small-scale landscapes, oriented as he is to the cultural landscapes of Britain rather than the vast North American spaces of the early earthworks artists (see, e.g., *Sheepfolds* 2002).

Carlson does not discuss Goldsworthy's artworks; we speculate that if he had, he would have argued that they aestheticize, even sentimentalize, nature. The use of color and form in many of Goldsworthy's works makes them highly artistic, even decorative in character, so a possible objection is that these works manipulate nature for aesthetic results, amounting to the aestheticization of nature. If one's only exposure to his impermanent work is through still photographic representations, such an objection might hold. However, an understanding of the processes, which are difficult to capture in stills, helps to address the objection. For example, the film about Goldsworthy's work, *Rivers and Tides* (Riedelsheimer 2004), supports an appreciation for the artistic process and struggles with nature inherent to his work, where there are many instances of failure among few successes. A closer look at this artist also shows that his interests are far from the decorative and pretty. He says, for example, that "colour for me is not pretty or decorative—it is raw with energy" (quoted in Matless and Revill 1995, 442). Furthermore, Matless and Revill also point to Goldsworthy's interest in decay, incongruity, contingency, indeterminacy, and erraticism. This isn't always evident in the ordered, final pieces we see in photographs of his works but is perhaps better understood through dialectical processes of creating art in conjunction with the unexpectedness, spontaneity, and uncontrollability of many natural materials and processes. Goldsworthy writes,

> Nature is intensely beautiful and at the same time very unnerving, and at times deeply frightening. You feel it as soon as you go out to the land, where everywhere you go things are dead, decaying, fallen down, growing, alive. There's this incredible vigour and energy and life. And it's sometimes very difficult to deal with. I would hope that I don't have a kind of romantic view of nature. I do feel the beauty of it, for sure. But it's a beauty that's underwritten by extreme feelings. (quoted in Tufnell 2006, 81)

Ben Tufnell takes this more sympathetic interpretation of Goldsworthy's art much further, drawing comparisons between his works and Smithson's, and claiming that Goldsworthy's work is "as obsessed with decay and disintegration as Smithson was" (2006, 81). This close comparison is not very persuasive, though, because the works of these artists are rather different in character. Although they are both interested in natural processes, Smithson's aesthetic is largely postindustrial and entropic, often on a larger scale, while Goldsworthy's focus is principally rural nature. In any case, understanding Goldsworthy's work in these ways makes it possible to more easily deflect the aestheticization objection.

Carlson's discussion of another artist supports our suspicions about his aesthetic affront objection, since he objects even in cases where the artistic relationship to nature is more intimate. Singer's artworks have been described by Auping as reflecting a "philosophy of extreme restraint and discretion towards landscape," which the artist himself describes as " 'research' in the form of a continuous dialogue with natural processes" (1983, 98). Despite apparent concessions to Singer on the grounds that his artworks highlight nature's qualities, Carlson's final conclusion is that these works may be seen as causing an aesthetic affront of another kind, "the affront implicit in the idea that for the aesthetic interest and merit of nature to be recognized it must first be considered a work of art" (2000, 161). In addition, he makes the further claim that because such works "do little if anything to alter nature's qualities . . . there are adequate grounds for considering these natural pieces not to be works of art in the first place," so they then drop out as potential cases for consideration in his argument (2000, 160–61).

Implicit in Carlson's claims is the "positive aesthetics thesis," which argues that "pristine" or "wild" nature is always beautiful or wild nature just is, always, "aesthetically good" (2000, 72). There are several problems with this position, but for our purposes here, the problem is that the thesis places the aesthetic value of more natural environments over cultural ones simply in virtue of natural value. Wild nature is just always more beautiful than humanly modified nature, and only the latter can ever have negative aesthetic value. Commenting on the claim that some artists improve on natural sites that would otherwise not have redeeming aesthetic qualities, Carlson writes,

> The desert, for example, has a subtle, quiet beauty of its own, and altering that beauty can be as great an aesthetic affront to nature as altering the aesthetic qualities of conventionally scenic landscapes. In fact, I suggest that none of virgin nature is comparable to the work of a third-rate hack—that virgin nature by and large has positive aesthetic qualities. (2000, 157)

This is true in so far as it shows why it would be absurd to assume that humans *necessarily* improve on nature, but there is an implicit prejudice in positive aesthetics against cultural landscapes and aesthetic alterations of the environment. By privileging the aesthetic value of wild over cultivated nature, it downplays the value of creative interaction with nature.

Positive aesthetics explains a short-sighted attitude toward exploring the variety of potentially rich human-nature relationships that emerge through art-making. It implies that artists can never add aesthetic value to nonmodified environments, and in its strongest form it holds that any artistic act in relation to wild nature constitutes aesthetic interference. Environmental harm is a serious concern but, as we argue, some modified environments express positive nature-human relationships where aesthetic experience and aesthetic value play a key role.

To get around the positive aesthetics objection, it might be helpful to conceive of some environmental artworks as comparable to gardens. In her philosophical and historical analysis of gardens, Ross argues that despite the decline of the art of gardening, "high art has not retreated from the landscape." Gardening is the "true ancestor" of environmental art, "environmental art is gardening's avant-garde . . . they address the relation work to site, like gardens; they can be ideological, like gardens; they can be beautiful or sublime, like gardens. Overall, they force us to think deeply about nature itself, about our relation to nature, and about nature's relation to art" (1998, 202, 204).

Granted, some types of gardens may be said to constitute an aesthetic affront, but if one accepts that many kinds of gardening represent a positive, even co-nurturing, relationship between humans and nature, then this may help to support the same claim in respect of some artworks. If the problem is approached in this way, it is possible to recognize how artists work more cooperatively with natural processes rather than simply using them as a means to their artistic ends. Like the gardener, many environmental artists must work with the forces of nature and, as such, these relationships are not necessarily conflicting or hierarchical; they can be two-way (as we have shown in earlier chapters).

Carlson's objection appears to be accusing some environmental artists of something like the same art-first view of nature held by picturesque theorists. Why should this be the case? Both the artists' aims (as gleaned from artists' statements), as well as the works themselves, suggest that they want to use artistic expression to work symbiotically with nature's qualities, rather than merely imposing an artistic vision onto them through intervention and alteration. For example, Singer has said of his work that "in order to experience and learn from the natural environment . . . I felt the need to yield to it, respect it, to observe, learn, and then work with it" (quoted in Saito 2002, 178). Goldsworthy concurs,

> I'm not an artist born full of things I want to express. I'm empty, hungry, wanting to know more. That's my true self; and my art is a way of learning, in which

> instincts guide me best. It is also very physical—I need the shock of touch, the resistance of place, materials and weather, the earth as my source. It is collaboration, a meeting point between my own and earth's nature. (quoted in Tufnell 2006, 78)

Here, Goldsworthy also points to the importance of understanding humans as both part of and, in many ways, different from nature.

Alan Sonfist, another target of Carlson's criticisms, is best known for *Time Landscape* (1965–2007), a restored precolonial forest ecosystem on the corner of a city block in New York. Although the materials and processes are predominantly natural, there is a clear artistic point, but it is not one that suggests an art-first kind of view. The project enables people to experience a stage in the history of the relationship between culture and nature in New York, a place that represents this "urban area's vanished native flora" (Ross 1998, 216). As an artwork that has been described by Ross as a "quasi-garden," it has also succeeded in providing a green space for the local community (although at present you cannot enter the space itself). In an interview, Sonfist relates his vision for *Time Landscape*:

> My art began in the street fires of the South Bronx, late 1950s, when I was a child. Gangs and packs of wild dogs were roaming the streets where I was growing up. The neighborhood was a landscape of concrete, no trees. The Bronx River divided the two major gangs, and the river protected a primal forest. It was my sanctuary as a child. The human violence didn't enter the forest; it was my magical cathedral. I would skip school to spend every moment I could in this forest and replenish my energy, my life. The forest became my life, and my art . . . The reconstructed forest was a way of going back into my childhood forest in New York as it would have been, initiated in Greenwich Village. I transplanted living tree species such as beech, oak, and maple and over two hundred different plant species native to New York, selected from a precolonial contact period in New York. These are still there on site. Besides experiencing the indigenous trees of New York City, Time Landscape© allowed me to experience and interact with foxes, deer, snakes, and eagles, and this was part of my experience. (Grande and Lucie-Smith 2004, 166)

Sonfist (as well as Singer) believes that nature can be valued without art as a mediator, but art can also mediate positive relationships between humans and nature. If such mediations support an aesthetic regard for nature, then they should be encouraged.

Time Landscape does not confuse art with nature; art mediates between the artistic agent and the environment in question. The works of Singer, Sonfist, and others are not reducible to natural environments as such. Compared to Heizer's earthworks and other sculptural pieces in the land, they lie on the more natural end of the scale; however, they may still be described as human-influenced and human-modified environments that emerge through interaction with natural processes and objects.

Ecological Art

We have so far discussed cases of environmental and land art, which have come under particular critical scrutiny. "Ecological art," on the other hand, presents the most obvious case of art that shows aesthetic regard for nature. Works of this kind tend to be driven by a biocentric environmental ethics and are often interested in bringing attention to human responsibility for the environmental crisis. Several artists may be included here: Agnes Denes, Helen Mayer Harrison and Newton Harrison, Patricia Johanson (and Sonfist too). These works often show direct concern for nature on a more global scale, with a strong interest in restoration and remediation of damaged ecology. Living and growing materials cultivated over time may be essential, as they create opportunities to interact deeply with natural processes beyond a single moment or static view.

Johanson's projects are remarkable for both their scale and success, as well as presenting early cases of remediation through art. She is clear that her aims are artistic: sculptural forms, colors, and the creation of beauty work in tandem with the science that will enable remediation, with significant outcomes, for example, better water quality, creating microhabitats, and restoring species. Johanson is interested in restoring local places that bring people and nature together, rather than restoration projects that would seek to keep people out in the name of preservation. Artistic and aesthetic qualities work to create interest and bring people into places such as *Fair Park Lagoon* (figure 5.2): "Fair Park Lagoon in Dallas . . . seeks a fusion of aesthetic form, functional infrastructure, and living ecology, where every element is part of a larger system whose parts are intricately related. The sculptural forms simultaneously control bank erosion, serve as paths and bridges over water and create microhabitats for a wide variety of plants, fish, turtles, and birds" (Kelley 2006, 25).

In *Petaluma Wetlands Park and Water Recycling Facility*, a more recent project (2001–2005) integrates water recycling with the restoration of a wetland to create a beautiful park used by the Petaluma community (Wu 2013, 166–76). Johanson's art presents a clear attempt to work harmoniously with nature and to make ecology central: "The most important aspect of my art is in the parts I do *not* design" (Kelley 2006, 19), and she wants this to go hand in hand with creating opportunities for meaningful relationships between people and place.

In such projects, we should not forget the role of nonhuman creatures. Insects and other tiny nonhuman creatures will play a role in many works,

Figure 5.2. Fair Park Lagoon, Dallas (1981–1986).

either intentionally or only incidentally, but a focus on larger nonhuman creatures such as amphibians, reptiles, birds, fish, and mammals is less common. Johanson's projects seek to attract creatures that form parts of the ecosystems she works to restore, but she is working more holistically rather than focusing on individual animals or species.

Bringing attention to huge numbers of extinctions among amphibians, self-described "artist/biologist/environmental activist" Ballengée used semi-domesticated varieties of the *Hymenochirus* frog, in *Species Re-clamation* (1999–2006), to selectively breed generations to produce a "wild-type" *Hymenochirus curtipes* that is nearly extinct (Ballengée 2017). Hull's trans-species art and habitat sculptures suggest that restitution motivates her work: "My sculpture and installations provide shelter, food, water or space for wildlife, as eco-atonement for their loss of habitat to human encroachment" (Hull 2009). For example, her series of *Raptor Roosts* (1988, 1990) provide safe roosting places for birds of prey susceptible to electrocution from power lines. Although situated in the land, her work provides a contrast to forms of land art discussed above:

> I've had a long-standing interest in earth or site-specific art, but too often it seemed so egocentric—on a grand scale, to go out and abuse the land in the name of art—which, as much as I love some of these pieces and enjoy hunting them down, did not seem enough. I felt a growing need to make some positive gesture to the earth. Couldn't there be a small-scale, nurturing, perhaps even "feminine" land art? (quoted in Gablik 2002, 89)

These ecological works use art practice to benefit nonhuman animals and, as such, can be seen as showing aesthetic regard for animals in particular, rather than thinking about ecological wholes.

CONCLUSION

This chapter has argued that artworks that show "aesthetic regard" capture an aesthetic relationship to environment that is more respectful of nature. As opposed to Carlson's preservationist approach, which largely denies the significance of artistic interaction and alteration of nature, we have also shown the potential of environmental art to provide insight into sensuous, imaginative, expressive, and conceptually based forms of human valuing of nature. These artistically generated relationships contribute a great deal to grasping the variety of ways that the natural environment is aesthetically valued, as well as indicating some connections between aesthetic and ethical value. In particular, we have seen that some forms of art enrich and extend our experiences of aesthetic value beyond those based in an "observational aesthetic," drawing attention to deeper forms of environmental engagement, including how natural processes unfold over time (Brady 2003). The complexity of aesthetic qualities in many environmental artworks also widens the range of aesthetic qualities that contrast with the beautiful, for example, ugliness and the unscenic, qualities that enable more balanced appreciation that is, arguably, more on nature's terms—however challenging it may be to understand those terms (Saito 1998). Our discussion of different forms of art and their connections to aesthetic regard is by no means exhaustive. The vast range of works in this area is daunting for philosophical analysis but, equally, intriguing and worth further study.

Chapter 6

Ecological Restoration

In the year 2000, some 2,500 people witnessed a group of ex-coal miners ascend the former Penallta colliery landscape in the Rhymney Valley, nestled within the South Wales Valleys. Led by the Penallta lodge banner and a brass band, the miners symbolically reclaimed the site that had stood as one of the largest deep coalmines in South Wales, before it closed in the early 1990s. This act formally reopened the site as Parc Penallta Community Park, after a four-year restoration project across the site, led by Caerphilly County Borough Council and Groundwork.[1] The site is now a 180-hectare community park, with different habitats (marshland, grassland, fishing lakes); new walking and bridle paths and cycle tracks; and public land art, including "Sultan" the Pit Pony, the United Kingdom's largest figurative earth sculpture. In this chapter, we describe the process by which the colliery was transformed into a park through restoration, focusing on the aesthetic values and qualities of the project.

The majority of ecological restoration scholarship within environmental philosophy has been conducted through a process of building *a priori* postulations about the types of values that restoration may be premised upon, and in turn the ontological status of nature and culture: are restorations "fakes" of originals (Elliot 1997)? Might restoration amount to another form of human domination over the nonhuman world (Katz 1993, 2000)? Is restoration a scientific endeavor or is it a type of artistic practice (Turner 1990; van Diggelen et al. 2001; Jordan III 2003)? By contrast, in this chapter, we are interested in the ways in which values have been conceptualized by restoration actors themselves, including how values were projected onto the landscape within policy documents, how these values were actually implemented, and the aesthetic qualities that have resulted.[2]

We demonstrate that aesthetic values intersect with other forms of natural and cultural value, including those pertaining to landscape and social heritage, ecology, and recreational landscape utility, and how this "intermixing" of aesthetic and other values (Berleant 1997, 62) brings about a particular set of aesthetic qualities we identify as being "populist," and an overall landscape character of functional beauty. Ultimately, we argue that in attending to both natural and cultural values, the project propagates aesthetic qualities and an overall landscape character that destabilizes such a binary; instead, they are clear aesthetic expressions of a dialectical relationship between the two. Before we do so, it is important to first consider the initial motivations to restore what became Parc Penallta and also the project's overarching objectives.

RESTORATION MOTIVATIONS AND OBJECTIVES

In 1983, the National Coal Board (NCB) was granted planning permission by the then Mid Glamorgan County Council (MGCC) to continue coal mining and on-site tipping of spoil, on the condition that a restoration and aftercare project would be undertaken once all coal activities ceased.[3] The colliery closed in 1991, tipping of spoil stopped in 1993, and coal stocking in 1994, by which point the NCB had become the British Coal Corporation (BCC) as a result of the privatization of the British coal industry. Rather than carry out the legal requirement for restoration themselves, BCC entered into negotiations to transfer the site to local government by means of a dowry that would cover the cost of this work. Protracted negotiations were concluded by mid-1996, with the dowry totaling £584,000—an amount that would have allowed only the new owners to undertake the legal minimum requirement, which was "turning it green essentially . . . they would have sprinkled grass seed on it and left it" (Paul Cooke 2009).

Around this time, Groundwork UK was putting in a bid to the Millennium Commission lottery grant stream for its Changing Places program, which resulted in £22.1 million in funding across the program's twenty-one sites. Groundwork Islwyn and Rhymney (which became part of Groundwork Caerphilly [GC] in 1996) approached Caerphilly County Borough Council (CCBC), proposing that it could access some of this funding so that "Groundwork could help the Local Authority to do something more than turn this tip green" by transforming the site in to a community park (Paul Cooke 2009). This raised £472,000, while an additional £1 million was subsequently forthcoming through the European Union's RECHAR (Objective 2) Programme, which is aimed at regenerating ex-coal mining areas. With a total amount of just over £2 million, "we're starting to talk about a pot of money then that we could do something with" (Lewis 2009).

The preexisting legal requirement to restore the coal spoil tip, once mining operations had ceased, meant that the NCB had already drawn up restoration plans before the site was transferred to the local government. These plans, approved by MGCC in October 1994, intended to "restore the tip as agricultural grassland, sub-divided into fields by fenced hedgerows, with copses planted to compliment areas of natural and previously restored woodland" (CCBC 1996, 2). No further detail is provided to ascertain whether the restoration of subdivided fields would retrace the exact historic spatial configurations of the agricultural landscape. However, the general premise reflects a desire to recast the landscape in such a way that the mining history is effectively attenuated—if not completely erased—from the landscape.

This reference point (or, to use ecological restoration parlance, the "reference model") was rejected by the CCBC and Groundwork restorationists, in part because of the practicalities of such a restoration: "We didn't pour over nineteenth century plans of the area which showed the original field patterns, because the bulk of the material tipped on that site would have made [their relaying] impractical . . . being on a slope" (Daniels 2009). However, even if this challenge of topography could have been overcome, this restoration strategy would still not have been implemented. Instead, a second reference model was brought forward that attempted to synergize cultural and ecological values, and, importantly, make these relations aesthetically legible, rather than fully erase the site's mining history through a process of naturalization.

This second model involved the melding of two different visions for the site. First, there was a desire—primarily driven by Groundwork's remit of engaging local communities within such projects—to preserve visible signs of coal mining activities, based upon an understanding by the policy makers that the coal landscape was positively valued in terms of what it *represented* to local communities, including the human labor that produced it: "we didn't want to wipe away the fact that it's a tip because it's part of the heritage round here, it's acknowledged that it's created by the mine" (Paul Cooke 2009).

Second—and seemingly at odds with this conservation of landform appearance—from the perspective of the CCBC landscape architects, the tip was viewed as in need of aesthetic *integration* into the landscape: "If you've got a bulk of material tipped on the valley side . . . you're going to be left with a landform that sticks out, unsightly, you're screening it in some way to then blend it back in with what's adjacent" (Daniels 2009). Indeed, prior to the restoration works, there was a consistent attribution of negative aesthetic value; for instance, policy documentation describes the site thus "an active railway line runs northwards up the valley into the storage areas, with disused sidings, embankments and a quarry to the west of Penallta road completing the picture of industrial dereliction" (Groundwork Islwyn and Rhymney 1995, 9). Further visual judgments of degradation were articulated during conversations with policy makers, who variously characterized the site as "just an old tip," "an eyesore," and a "huge black scar," which was prone to attracting fly-tipping.[4] This second vision, then, does not relate to a concern for diachronic integrity, which is an attendance to the continuation of a narrative of a place through time (Holland and O'Neill 1998); rather, it speaks of the potential for *spatial* integrity. We must keep in mind that some form of ecological restoration—however tenuously restorative—is a legal obligation, and so some form of "greening" the black tip must take place. Arguably, then, this legal obligation will inherently reduce the visible progeny of the site's landform and increase the chances that the tip will "blend back in."

Nonetheless, what is proposed goes beyond passively "greening" the site: the assessment of the tip's existing aesthetic qualities led to a perceived imperative to restore the site, using the visual aesthetics of the present-day agricultural landscapes located adjacent to the site as a reference guide, which is "valley side woodland, open field, [and] wetland corridor" (Daniels 2009). Such "different place, same time" restoration models (White and Walker 1997, 345) are often used when historical information of a site is incomplete or nonexistent. Yet, in this case, there were no attempts to draw on the pre-coal mining agricultural landscape of the late eighteenth century as a reference guide, either from a spatial perspective—for example, to look at where specific landscape formations should be located with regard to one another—or from a species perspective. In fact, there is acknowledgment that species that would not have been present at this time, such as nonnative Italian Alder (*Alnus cordata*), will be planted because of the difficulty of floral establishment on bare shale substrate and the subsequent "visual diversity" that this planting will bring to the site (Daniels 2009). Additionally, only the visual *appearance* of the adjacent reference sites will act as a guide; the nutrient quality of the grasses on the tip will be too low to support animal grazing, and artificial fertilization will be avoided (Daniels 2009), meaning that post-restoration the site will not support agriculture.

The melding of these two different visions led to the production of the following mission statement:

> To develop a Community Park on 180 hectares of derelict and underused land at Penallta. To improve the quality of life and the community pride of people around the former colliery at Penallta by developing community participation in environmental improvements and management. (Groundwork Caerphilly 1996, 5)

We see then that the central aim of the restoration has widened in scope: the original legal conditionality of reclaiming the coal tip now encapsulates a strategy of postindustrial *cultural* regeneration, with human "communities" being the benefactors of such restoration.[5]

A "seven-strand programme" of objectives was drawn up to implement this mission statement. These were the following: (1) community involvement, including various rounds of local consultation on the restoration proposals; (2) land reclamation through earthworks and drainage, and the restoration of land to productive use; (3) the provision of facilities for formal and informal recreational activities; (4) an improvement in access through a network of foot and cycle paths and the development of links to the wider countryside; (5) nature conservation by managing existing and developing new habitats; (6) the preservation of cultural and heritage features; and (7) the development of a community art program (Groundwork Islwyn and Rhymney 1995, 4–5).

While the original legal requirement for restoration was applicable to the Penallta coal tip and part of Nelson Bog SSSI,[6] the proposed park incorporated adjacent areas already under CCBC ownership, including a sandstone outcrop known as Penallta Rocks; Coed Penallta, an adjoining small area of woodland; and Penallta Marsh, a small area of boggy grassland that is being sensitively managed in an attempt to restore nationally declining marsh fritillary butterflies (*Euphydryas aurinia*). The inclusion of these areas brought about benefits, both in terms of strategically consolidating landscape management practices and because it was seen to increase the landscape's ecological and visual aesthetic value when recast as a park: "We sort of expanded the size of the park, so the park wasn't then just a tip site . . . it was actually some sort of really stunning natural scenery and important natural habitat as well, so we can . . . pull it all together [and] look after it as a whole" (Paul Cooke 2009).

POPULIST AESTHETIC VALUES THROUGH LANDSCAPE DESIGN

We have seen that, far from merely "greening" the coal tip, as was legally required, the restorationists from both GC and CCBC drew up a restoration strategy that placed particular emphasis on the role of landscape heritage and community arts, as well as opening up the landscape for recreational use and physical access; this was in addition to the more common ecological restoration practice of landscape reclamation. The realization of this strategy, we suggest, has given rise to a form of landscape aesthetics that we identify as "populist." In line with how the term has been deployed in architectural theory (Stead 2004; Kühn 2005), we use it as a means to describe the creation of landscape features and aesthetic qualities that are aimed to appeal to a wide range of potential users and a set of landscape design motifs that are literal (as opposed to conceptual), easy to read and interpret, and speak to the site's recent working history. We will now outline a selection of these landscape features and design motifs.

Flat and expansive grassland, with some gentle inclines, dominates the top half of the park, which sits at a higher elevation than the lower half. Some patches are dense and scrubby with abundant clover, common daisy, and other ground-covering plants. Others are closely cropped and thin, revealing the coal shale substrate underneath the vegetation. The eastern portion of this landscape has been demarcated as an "events area"; this can accommodate up to 3,000 people, having done so during a kite-flying festival in the autumn of 2007 (Lewis 2009). Due to the open nature of this landscape, "the winds are quite bad so the engineers recommended that we create [a] wall if you like to stop the wind blowing everyone away basically" (Lewis 2009). Groundwork tendered out the work to produce this sheltering wall, which was awarded to the Wales-based landscape designer and artist Mick Petts. The resulting wall took the form of a figurative earth sculpture called "Sultan the Pit Pony"; this was sculpted directly out of the coal shale and at 200 meters in length remains as one of the largest earth sculptures in Europe (Lewis 2009).

Sultan, as with many of the designed motifs across the park, is a literal expression of the near-recent coal mining activities at the site. Prior to mechanization, ponies were used to haul coal underground. The ponies would spend their whole working lives in the mine, except for two weeks each year when the pits were closed for the miners to take a holiday; during this time, the ponies were free to roam above ground. The sculpture takes its name from a pony that worked at Penallta colliery; so the story goes, Sultan was a winner of many local pony shows (Lewis 2009). Once excavated out of the shale, Sultan was hydroseeded: this is a method of substrate stabilization, where a mixture of sewage sludge, grass, and wildflower seed is sprayed directly onto the shale (Lewis 2009). The wildflower mix was nonnative in species composition; instead, species used were selected for their visual qualities (Daniels 2009). The "eye" of Sultan is shaped out of anthracite coal that, ironically, had to be collected from an active mine in Swansea as all local mines are now closed. Mowing and strimming is carried out on an irregular basis across the grassland, in response to continued observations and also budgetary constraints, resulting in temporally uneven aesthetic qualities: premanagement, Sultan has a "scraggy" appearance with a diversity of wild flowers; post-management, it is more akin to a "lean race horse" (Lewis 2009) with clean lines.

Located to the east of Sultan and the events area is a hill colloquially known as "Yellow Mountain" due to the presence of grasses that "would shimmer in the wind . . . like a yellow haze" (Lewis 2009). The restorationists wanted to put in place a sculptural form that would attract people to the top of the hill and act as a visual beacon from the outlying landscape. Once again, a competitive tender process resulted in the creation of an observatory sculpture by Scottish artist Malcolm Robinson. Made from stainless and weathering steel, the sculpture is shaped as a compass extending out into the landscape. Each arm of the sculpture is directed toward a distinct landmark, such as Llancaiach Fawr, a sixteenth-century manor house, distant church spires, and the adjacent Nelson Bog. Schoolchildren produced drawings to represent these features, which were subsequently laser-cut into each relevant arm of the sculpture by Malcolm.

At the north foot of the hill is a sequence of twelve low-lying stone installations entitled "Stone Stories." The Wales-based sculptor Robert Kennedy conducted a series of workshops and interviews with ex-miners "to find out what they were up to [when mining] and what they did and thought about the site" (Lewis 2009). Excerpts from these sessions were sand blasted onto twelve pennant stones that double as informal seating. Situated near to the main entrance to the park is a piece by the artist Godfrey Phillips called "Coal Cutters Dream," carved from the trunk of a locally felled cedar tree (Phillips 2003, 1). This depicts a coal miner sitting down leaning against his coal-cutting machinery, while dreaming about being up on the surface of the landscape (Lewis 2009).

To the west of Sultan is a large flat area of scrub grass that is of generally poor vegetation cover and is boggy underfoot during periods of rain. This area was purposefully not completely grassed over—unlike the events area—for ecological reasons. The area is "potentially good for lapwings,"[7] which require open, bare land for breeding, and no overhead vegetation in which predatory carrion crows can hide, so any developing tree or shrub species are removed (Lewis 2009). Diggers have been used to hollow out shallow pools in the shape of pony hoof imprints, which now attract a range of dragonfly, newt, and frog species (Lewis 2009).

A variety of other aquatic designs have been implemented across the park, having arisen to serve different landscape purposes, and because, from a sensory perspective, water "adds interest" (Daniels 2009). A wetland area has been excavated adjacent to Nelson Bog SSSI in what was a previously "barren" area: "we wanted to kind of mirror what habitat you might find in Nelson Bog, without attracting people in to quite a dangerous habitat and quite a fragile habitat" (Lewis 2009). To produce this experiential mirror of Nelson Bog, bulrush, iris, and other plant species were harvested from the bog and planted within the wetland. Access across the wetland has been made possible through the provision of a wooden boardwalk (figure 6.1).

Figure 6.1. View of the excavated wetland from the wooden boardwalk. Note the wooden pit props that are part of the boardwalk design and also the bare shale substrate on the incline in the background.

A series of wooden pit props have been incorporated in to the design; these are structures used to support the ceiling of excavated mine tunnels, so again we see literal sculptural echoes of the subterranean colliery landscape.

Lying to the east of the wetland area are two coarse fishing lakes. These were initiated through consultation with local angling clubs, overseen by the Rhymney Valley Federation of Angling Clubs. Originally, these were four settling lagoons that filtered silt that had been washed off of the tip on higher ground, but these were deemed "ugly" and "dangerous" for the park setting (Daniels 2009; Lewis 2009).[8] Benches have been provided adjacent to both lakes, as have fishing platforms for ease of access to the water's edge.

Continuing southeast of the fishing lakes is Penallta Rocks, a sandstone outcrop categorized as a Site of Importance for Nature Conservation (SINC). This affords the Rocks limited preservation and as such does not preclude human engagement; indeed, they are well used for rock climbing activities. The surrounding predominantly larch woodland preexisted the project, having been planted by the Forestry Commission for commercial timber production. While the majority of the park is open and expansive, this woodland landscape presented the opportunity for the restorationists to create an intimate, enclosed experience for visitors (Daniels 2009).

In 2000, a document was produced that attempted to create a coherent arts strategy for the entire park. Focusing on the theme of "energy in the landscape," plans were drawn up to create

> artworks and structures [to] interpret energy in all of its contemporary forms. These would include coal, wind, water potential and solar energy. These features would work to add visual and sensory interest to the Park, as well as providing the backbone for extensive use within a developing education programme. (Petts and Dale 2000, 1)

Within the document, forty-nine different artworks and structures were envisaged on the theme of energy. For instance, one of the proposed artworks was a horizontal visual representation of the vertical layers through Penallta's geology; samples of rock and coal would act as markers, while sound posts would be used to play "descriptions recorded from former miners of what it was like down there . . . or sound pictures created to convey what it sounded like" (Petts and Dale 2000, 2). Another proposed piece would have seen the

Figure 6.2. A footbridge over a stone drainage gulley links the higher and lower sections of the park. The stepped pathway to the left of the image brings you to the large flat area of scrub grass to the west of Sultan, while the path to the right leads to the wetland area and fishing ponds. The landscape in the background of the photograph, which lies outside of the perimeter of the park, demonstrates the "valley side woodland, open field" reference guide.

addition of a listening post within Sultan the Pit Pony's ear, playing "short stories, poems and sound images of the Pit Pony's life thousands of feet below ground" (Petts and Dale 2000, 3), while another would have involved the creation of interventions within drainage gullies that have been created throughout the park (see figure 6.2): small weirs would allow water to flow in to stepped pools, and an installed sound system would make "the delicate and varying sounds of trickling, gurgling water more audible . . . attracting visitors to investigate" (Petts and Dale 2000, 5).

At this time, "there was some talk of it [Parc Penallta] being an art park more than maybe a Country Park" (Daniels 2009), but this was eventually rejected, meaning that the ambitions of the arts strategy document were severely scaled back, including scrapping all of the proposed sound art works. In addition to those artworks already documented, a solitary brown hare oak sculpture was sited on the northern slope of Yellow Mountain, while a sculpture of a seed emerging from a concrete block titled "Breaking the mould" by Andrew McKeown sits adjacent to the events area, but the same sculpture is on display at all twenty-one Groundwork Changing Places sites. The reasoning behind rejecting the art park vision is summarized as follows:

> I didn't think that the authority overall, if we had quality art there, and there's plenty of poor quality art about, I felt that we should be insuring it and taking a liability and I didn't think that perhaps we were ready for that . . . we didn't want it to turn in to a Country Park which is stately house grounds if you like, full of some very expensive pieces of art work. (Daniels 2009)

Here we see the rejection of the art park concept due to concerns about the maintenance of art works within the landscape but also for its assumed aesthetic and affective repercussions. "Expensive pieces of art work," in the mind's eye of the park's landscape architect, would have necessitated the provision—or helped in the production of—"stately house grounds." Such a landscape transformation has clear connotations of elitism at odds with the overriding appeal to populism, which is instead better served through a "low-key" "comfortable" form of accessible countryside (Daniels 2009).

Through the art works selected, we see the conscious production of a variety of visual, textual, and textural[9] representations of the site's recent coal mining history. This process, with nominal local community input but maximal appeal to that community, results in artworks that are *selectively* populist. For example, no history of the premining agricultural landscape is represented at the park. Additionally, those representations chosen are a nostalgic—even whimsical—agglomeration of mining tropes: pit props, ponies, and resting miners. While this selectivity came about in a rather top-down manner, the artworks *do* make an appeal to popular memory. They act, then, as a populist memorialization of coal mining and the surrounding communities or, at the least, as a form of populist "guided remembrance" (Kearns et al. 2010, 734), and amount to what we may term the "heritagization" of the mining landscape (Conesa et al. 2008; Ruiz-Ballesteros et al. 2009), to achieve the symbolic representation of the site's former use.

There are dangers with such heritagization strategies, as Walsh suggests,

> It is of course arguable that the heritagization of space can help maintain an identity of place, through the emphasis on historical characteristics which stand as a metaphor for that place. The preservation of such images may be all the more important as local industries and communities are destroyed. The danger is, however, that only safe and selected images will be preserved, and the history of place will be neglected, while the heritage, over subsequent generations, helps construct an image of place which is based on superficialities. (1992, 139)

While we cannot know how subsequent generations will interpret these pieces, to write off the act of *in-situ* landscape representation as a consumable spectacle (Walsh 1992, 139) would at the least be an inaccurate reflection of the restorationists' intentions. Additionally, this is less about an act of grandiose territorialization through monument building, than is the case with many governmental and private-sponsored heritage projects (McNeill and Tewdwr-Jones 2003; Breglia 2006). The intention is for the art works to act as a mnemonic device (Boyer 1994, 343) that speaks to and for a "vernacular heritage" that is "particular and lived rather than general and abstract" (Healy 2001, 280). Rather than "safe" and "superficial," this form of heritagization affords the opportunity for the aesthetic representation of lived experience, inculcating affective qualities of remembrance and celebration of workers and the worked landscape.

LANDSCAPE MOVEMENT

In addition to these various landscape features that attest to a literal, populist, representation of the site's recent history, there was a concerted effort to also afford populist aesthetic *encounters* across the transformed landscape. The overwhelming design objective here was to instill a sense of comfort within park visitors, in terms of ease of comprehension of the park and its navigability (Daniels 2009). This is attended to immediately upon entering the park. The main entry point to the park brings you to a small wooden building that serves as a visitor center. From here, maps of the park are available for free, alongside other material relevant to Penallta and the surrounding area. A park warden's office is located within the center, on-hand to receive school groups or general visitors and to provide advice, information, and directions.

When the center is shut, maps of the park are still available from a dispenser attached to the side of the building.

From the car park are three different circular walking routes internal to the park, which are clearly plotted on the map and are signposted across the site. These are differentiated according to how "difficult" they are, both in terms of their distance and the gradient of the land (Daniels 2009). The shortest route is the "Sultan Trail," which circles past Sultan the Pit Pony and the events area and loops below the foot of Yellow Mountain. The "Skylark Trail" also follows the belly of Sultan, but curves round to the east, running around the area that has been excavated with pony hoof imprints. Interestingly, the route does not correspond with the area that has been identified as the best place to see or hear skylarks (*Alauda arvensis*), which is on Yellow Mountain (Lewis 2009).

Finally, the "Woodpecker Trail," by far the longest planned route, extends down to the lower section of the park, running adjacent to the wetland area. It should be noted, however, that these routes allow for flexibility in walking directions either on or off pathways. The ability to walk off-route or even abandon a circular walk is intentional: "[the walks] should have a start and a finishing point, there should always be the opportunity to go from A to B but diverting to go back to where you started if suddenly it's too much for you, or if you've forgotten to turn the gas off or something" (Daniels 2009). Pathways are cleared of any encroaching vegetation to maintain uninhibited mobility across the site (figure 6.2), and there are plans to gradually replace the existing gravel pathways with tarmac to provide a smooth surface for easier, more efficient, mobility across a range of gradients, which has already been carried out on a path adjacent to the fishing ponds.

One of the challenges for the restorationists has been to find a way to find peripatetic linkages between the higher and lower sections of the park:

> You get a lot of people going to the top of the site, but it's hard to get people going right round the site because it's a long walk, and a steep walk, and similarly we get a lot of people going through the bottom of the site along the cycleway, who may never know what's at the top of the site. (Daniels 2009)

Though money has been lacking to implement it, a cable car running between the top and bottom sections of the site has been proposed as a potential remedy (Daniels 2009). Due to the walking distance of the wetland and pond areas from the park's main entrance, a small dipping pond has been hollowed out adjacent to the visitor center; this was designed "specifically for school groups . . . so they didn't have to be Landrovered down to the bottom of the site, or walked down to the bottom of the site, then Landrovered back up" (Daniels 2009). Across the park, there were also efforts to widen mobility access for other users. A bridleway has been provided following a section of the Woodpecker Trail, while a section of the National Cycle

Route number 47,[10] which stretches across south Wales from Newport to Fishguard, traverses the southern boundary of the park past Nelson Bog and the fishing lakes.

While we see that there is a clear focus on the movement of people, equally important is the demarcation of areas for physical rest and landscape contemplation. Along the different traveling routes are strategically placed formal seating arrangements, including picnic tables and benches. These are not contained to any one particular area of the park. Rather, they are sited at major path junctions (figure 6.2), adjacent to recreational activity sites, and within more secluded areas. Additionally, due to the expanses of smooth ground and the design of some of the sculptural forms, there are also a number of informal seating opportunities. The difference in gradient across the park, while presenting a challenge to the circulation of people, has allowed for the installation of a series of designated viewing areas for scenic landscape consumption. The previously mentioned observatory sculpture encourages 360-degree viewing of landscapes external to the park, while a second to the south of this affords a view of Ystrad Mynach. Two other viewing platforms have been designed to direct visual attention to designed features within the park, namely Sultan the Pit Pony and the wetland area. These latter two appear to be less about creating static pictorial scenes for landscape appreciation—a commonly voiced criticism within the environmental aesthetics literature—and more to do with allowing the full comprehension of landscape elements that cannot be gained at close proximity.[11] Attention to internal coherency is provided through consistency in signage and materials used (primarily similar types of wood and stone), and the repeated use of a "wave" motif throughout the park on signposts, fencing, and bridges (figure 6.2), which came about because "basically it's always bloody windy up here, so everything's always waving and flowing in the wind" (Lewis 2009).

ECOLOGICAL RESTORATION BETWEEN NATURE AND CULTURE

Ecological restoration, as it is mostly commonly conceived, tends toward the reintroduction of historical landscape states. For one, Throop (2000) considers ecological restoration to "involve minimizing some prior human impact by restoring a system to a healthier state, typically a state involving the species and processes that characterized the predisturbance system" (Throop 2000, 11). A historical point in time is chosen as a reference point for a restoration, and the natural landscape qualities and features present at this time act as a blueprint for restoration activities: animal and plant species are reintroduced; soils are detoxified; infrastructure, fences, and roads are removed; rivers are remeandered; and so on.

We have seen, however, that in the making of Parc Penallta, any sense of replicating an historical state has not been pursued. As a consequence, we see an embrace (consciously or not) of an alternative goal to the complete restoration of historical systems—so-called "future-oriented" restoration—that has been championed by various restoration theorists. For example, Hobbs and Harris (2001) believe that, while still restoring ecological function to degraded ecosystems, restorationists should be "setting goals for restoration which focus on the desired characteristics for the system *in the future*" (Hobbs and Harris 2001, 241; emphasis in original; see also Higgs et al. 2014).

At Parc Penallta, there has not been a uniform process of naturalizing the site either. To be sure, there have been efforts to locate the park as a "gateway" to the wider countryside: "We wanted to encourage people if you like their first step in to the countryside on ground that they were comfortable with, i.e., a park setting, and then encourage them to go further and explore the remaining countryside" (Daniels 2009). The manner in which the tip has been visually integrated into the surrounding landscape plays a significant role in the proper functioning of the park-as-gateway. Using adjacent landscapes as a guide for restoration allows for the inscription of visible "countryside" tropes—valley side woodland and open fields—that reflect the types of aesthetic qualities that will be experienced when visitors explore the "remaining countryside." Nonetheless, the project is not based upon a restoration narrative of "returning to origins" (Potteiger and Purinton 1998, 226) that attempts to realize what Carter Park and Coppack (1994, 166) refer to as a "contrived" form of rurality that is "unspoiled, unsullied and pasteurized."

Rather than extricating complex human/nonhuman interactions from one another, we see a form of landscape design that centers on the interweaving of biotic and abiotic nonhuman nature at various levels of spatial organization (grasslands, woodlands, wetlands, particular animal and plant species, rock formations, coal substrate, nutrient and hydrological cycles), with human systems and practices of political economy, aesthetics, cultural heritage, nature preservation, resource extraction, and recreation. The park, then, exhibits a tension between urban fringe and rural countryside, never fully inhabiting either category. This goes against much spatial planning that is enacted at the perimeter of urban spaces, which generally "aims at regulation and the creation of places with an unquestionable character of city or country, nature or culture" (Qviström 2007, 270; see also Edensor 2005, 54). This binary production of space is also pervasive in environmental policy, where it serves to delineate spaces of preservation as oppositional to spaces of degradation (Zimmerer 2000, 362–63), and reinforces the conceptual boundary between "nature" and the "urban," including in urban park settings (Gabriel 2011).

Because of the explicit synthesis of nature and culture in the making of the park, and because of the focus on populist landscape utility and movement, we see the emergence of an aesthetic character of "functional beauty," rather than naturalistic or wild aesthetic qualities (Brady and Prior 2017). Parsons and Carlson (2008) outline "functional beauty" as "a thing's aesthetic qualities emerging from its function or something closely related to its function, such as its purpose, use, or end" (Parsons and Carlson 2008, 2). However, given the specific design interventions within the park that we have outlined, it seems necessary to modify this understanding of the relationship between aesthetic qualities and function, from only the former emerging *from* the latter, to the former emerging *alongside* the latter.

Consider, for example, the grassland landscape that has been explicitly sculpted to function as a site within which to stage large-scale events. This landscape is managed to prevent vegetal growth from inhibiting its utility. From this management emerges visual qualities of clean lines, near-uniform coloration, and general orderliness, and so an aesthetic quality of "simplicity," which, when understood in the context of a functional landscape, can be considered as a type of functional beauty (Parsons and Carlson 2008, 96), rather than, say, a boring or mundane landscape. Yet, we can also consider this simplicity as a *precursor* to the landscape successfully functioning as an events area. Thus, both the aesthetic qualities and function of the landscape necessarily emerge in tandem with one another. Broadening out our focus, the park taken as a whole has been designed, and is being actively managed, with clarity and coherence for human users in mind. When understood in the context of its function as a space for easily navigable and safe utility, the park demonstrates both mechanical function, where we see efficiency, exactness, and repetition of patterns[12] (Berleant 1997, 88), and organic function, where we see integrity and harmony between elements (Berleant 1997, 88–89)—that is, not only the various spaces that comprise the park, but also the park in relation to the wider landscape.

CONCLUSION

Two different visions of the future Parc Penallta landscape were synthesized into the landscape reference model: diachronic integrity, in which the structural composition of the tip would be maintained due to its postindustrial heritage value, and spatial integrity, in which the tip would be integrated within the surrounding landscape, using a landscape character of "valley side woodland, open field, and wetland corridor" as a visual design guide. These two seemingly contradictory visions negated any sense that the pre-coal mining landscape of the late eighteenth century should serve as the reference guide for the project, as was planned by the NCB in 1994; instead, the meshing of these two visions necessarily created a "future-oriented" vision for the site.

As we went on to show, this was accomplished through a mixture of landscape preservation, the introduction of "new" habitats, and a heritage arts strategy. These were implemented on the basis of a landscape design aesthetic we identified here as "populist," in that it: (a) seeks to appeal to a wide range of recreational forms of utility; (b) speaks to vernacular heritage associations that are literal rather than abstract (pit props, pit ponies, pony hoof imprints, resting coal miners); (c) encourages various forms of landscape mobility (walking, cycling, horse riding) in a comfortable and easily navigable setting; and (d) stages a variety of aesthetic encounters, from the intimate to the scenic, while maintaining an internal aesthetic coherency that is easy to comprehend. We then explored how through this form of landscape design we see the emergence of functional beauty as an overarching aesthetic character, which at Parc Penallta includes both mechanical and organic functions, and importantly how this emergence occurred in conjunction with the project's focus on recreational utility.

We have also demonstrated how the means by which these two visions were synthesized within the reference model, and the resulting aesthetic qualities and landscape character, reveal much about the environmental and cultural ethical values of the project. The restorationists have used the project as an opportunity to identify both potential and existing ecological values, which led to attempts to restore certain animal species, including marsh fritillary butterflies on Penallta Marsh and lapwings and dragonflies on the grasslands adjacent to Sultan, and also conserve preexisting habitats and geological features primarily through expanding the boundary of the park.

The park has also been designed in such a manner that it echoes the adjacent countryside and acts as a "comfortable" conduit to it. This was done, however, in a way that resisted the full spatial aesthetic integration of the park within the wider landscape. Indeed, the restorationists rejected the total naturalization of the cultural landscape through popular artistic representations of coal mining, their disinclination to conceal the coal substrate from which the park has been carved, the manner in which the park was officially opened, and the conscious production of functional—rather than naturalistic or wild—beauty. As a result, we see the representation of a popular cultural landscape politics that is markedly different to the cultivation of "elitist" aesthetic qualities (Soper 2003) that "obscure . . . not only the forces and relations of production but also more plebeian, less pictorial, experiences of nature" (Daniels 1989, 206), as has been the case with other postindustrial urban restoration projects (Foster 2005). Aesthetics, then, becomes a means to project desired dialectical relations between nature and culture, to the point where aesthetics and ethics at Parc Penallta are inseparable.

NOTES

1. Groundwork is a public-private initiative, working under the rubric of sustainable development, through funding from the UK government, the European Union, the National Lottery, and the private sector.
2. What we describe is based upon interviews with key policy makers involved in the project, alongside an analysis of policy documentation, undertaken during 2009 (Prior 2012).
3. "Spoil" is the rocky waste material—predominantly shale at Parc Penallta—extracted as a by-product of coal mining.
4. Fly-tipping is the unsanctioned discarding of waste materials.
5. It should be noted that "community" refers to eight settlements adjacent to the Penallta colliery site (Cefn Hengoed, Gelligaer, Hengoed, Maesycwmmer, Nelson, Penpedairheol, Penybryn, Ystrad Mynach).
6. Nelson Bog is a raised bog that has been designated as a Site of Special Scientific Interest (SSSI), which affords it some conservation protection.
7. Lapwings (*Vanellus vanellus*) are wader birds that have significantly declined in number across Wales since the mid-twentieth century.
8. The importance of water drainage engineering is of vital importance to the stability, and thus the safety, of coal tips, which has been "learnt from past disasters" (Lewis 2009). Of particular pertinence to Penallta is the 1966 Aberfan disaster. Heavy rain caused a tip to subside and slide down a mountainside on to the village school and about twenty houses, killing 144 people, of which 116 were schoolchildren (Mclean 1997, 285–86). Aberfan is located about seven miles northwest of Penallta, and the disaster is still an emotive topic in the local area (Paul Cooke 2009).
9. The Parc Penallta artworks have been produced out of "big heavy" and "chunky" building materials, in part not only to reduce the effects of weathering, but also to reflect "the heavy industry within the area" (Daniels 2009).
10. This is a part of the United Kingdom's National Cycle Network of cycling routes.
11. See chapter 2 for a discussion of the "picturing" of landscape scenes and the picturesque.
12. For example, the repeating wave pattern on signage, fences, and bridges.

Chapter 7

Conclusion

Aesthetics and Ethics between Nature and Culture

In this concluding chapter, we work toward an understanding of the particular contribution that thinking about aesthetics between nature and culture can make to discussions about the intersections between environmental aesthetics and ethics. We have made some progress already in understanding how positive and meaningful human-nature relationships develop through various practices such as appreciating gardens or making land art, but here we bring our thinking more closely into dialogue with debates in environmental aesthetics and ethics. Let's begin by reviewing some of the key arguments of the book.

In the first two chapters, we explored gardens as a meeting point of humans and nature and focused on two challenging cases, the picturesque and topiary. The picturesque is an aesthetic theory and landscape design practice, which in turn shapes a form of aesthetic appreciation. We showed that while the picturesque has its limitations as an anthropocentric approach to the natural world, it does allow for appreciation of wild nature within the confines of a particular and historical conception of landscape. The picturesque provides an early example of how people form relationships with wild nature on terms that were possible at the time, as well as showing how human-nature relationships emerge within the context of gardens and designed landscapes. In both its historical and contemporary practice, topiary is a challenging case because it can involve the strong manipulation of plants by gardeners. We showed that some topiary practices reveal a caring relationship that aims to work more harmoniously with the growth of shrubs and enable their flourishing, if only within the context of a more formal garden style.

Moving out of the garden and into more expansive modified environments, we considered the tensions and harmonies, respectively, that emerge in the generation and aesthetic experience of agriculture and environmental, ecological, and land art. We argued that practices such as hedge-laying, coppicing, and stone-walling, which involve closer interactions with the land, present opportunities to develop more harmonious relationships and more intimate forms of aesthetic engagement within a context of production rather than explicit artistic creation. We underlined the importance of not romanticizing these practices, however, as more "traditional" practices can hide historical injustices. Some practices with attractive aesthetic outcomes, say, a beautiful stone wall, may be linked historically to the oppression of landless people by a handful of wealthy landowners. In the present day, a tension exists between the hardship and economic fragility of sheep farming in places like the English Lake District and the biodiversity desert linked to that practice (Monbiot 2013; Brook 2018). We found that care is also needed when considering the practices of artists shaping the landscape, whether they identify themselves as sculptors shaping the material of landscape (e.g., Michael Heizer) or as ecological artists (e.g., Patricia Johanson) intent on putting natural processes first in the remediation of ecosystems degraded through human activity. By showing "aesthetic regard" for nature in their work, we argued that some practices embody a relationship that is more respectful.

The theme of remediation for past damage to the environment is a strong motivation behind ecological restoration projects. In the penultimate chapter of the book, using the case study of Parc Penallta in Wales, we reflected on the place of aesthetic values in the design and reception of ecologically restored landscapes. The making of Parc Penallta illustrates the breaking down of human-nature binaries through an inclusive future-oriented vision that succeeded in interweaving cultural and natural heritage. It is fitting that a community park should serve as the final case study in our discussion of aesthetics between nature and culture, for it so nicely encapsulates our interest in how aesthetics features in landscapes that are the meeting points of people and place.

Throughout the book, we have been interested in how different forms of aesthetic production and appreciation foster—or not—harmonious relationships. On the one hand, our interest in this topic has been motivated by the relative neglect of modified environments within environmental aesthetics. On the other hand, we have been inspired by alternatives to intrinsic value approaches in environmental ethics, for example, environmental pragmatism and its attention to pluralistic and practice-based thinking, as well as ideas concerning meaningful relationships and notions of establishing "bonds" with the natural world (Light and Katz 1996; O'Neill et al.; James 2016). In light of this, our starting point for thinking about how aesthetics and ethics interact in the context of modified environments has been relationships. In order to characterize interactions between nature and culture, we have drawn upon concepts such as care, respect, aesthetic regard, and remediation, as well as exploring more problematic interactions such as aesthetic affronts against nature and conflicting relationships. We think that this approach has several advantages that advance thinking about the intersection between aesthetics and ethics.

In contemporary aesthetics, discussions about the relationship between aesthetic and ethical value take various forms, but they have tended to focus on whether or not there is an internal relation between a work of art's moral character and its artistic value. In environmental aesthetics, the debate has focused more on whether and to what extent aesthetic experience and value may be said to support moral attitudes toward the environment (Brady 2003; Carlson and Lintott 2007; Carlson 2010). For example, "aesthetic preservationism" holds that through the sensitive perception characteristic of aesthetic attention and the discovery of beauty, majesty, and so on, we may develop care and respect for nature. On this view, a particular kind of aesthetic awareness potentially feeds into ethical attitudes and forms of environmental action (Hargrove 1989; Thompson 1995; Rolston 2002; Fisher 2003). A version of aesthetic preservationism, Hettinger's (2008) "aesthetic protectionism," argues that valuing natural beauty can serve as an important motivation for protecting the environment, as long as sufficient justification can be provided—some kind of objectivity—for aesthetic judgments of nature. To take an example with which many people will be familiar, the destruction of a natural area for some kind of human development—industry, mineral extraction, housing, and so forth—will mean the loss of plant life and habitat for insects, birds, and other animals and, with that, changes and losses in the aesthetic qualities and character of a place. In anticipation of the possibility of this kind of loss, many people will seek to protect the area for ecological, aesthetic, and other reasons. Here, the valuing of aesthetic qualities may be cited as a motivating factor in any moral actions. The focus of aesthetic preservationism is, generally, on the intrinsic valuing of wilder environments, and it is intended to parallel environmental preservationism that, historically, has focused on wilderness areas.

We agree that finding aesthetic value in nature can potentially motivate environmental protection, but we think it is also important to draw attention to the kinds of relationships that develop within modified environments. Our approach provides a valuable addition to thinking about how aesthetics supports environmental ethics, and although it is not the only alternative to aesthetic preservationism, it is the first, to our knowledge, that strongly emphasizes the role of relationships.[1]

Among the advantages of our position is that it foregrounds the entangling of humans and environment. Understanding humans and more-than-humans as entities interconnected through a web of relationships helps to support an idea of community rather than human domination of nature and other hierarchical dualisms that have underpinned historical and deeply anthropocentric thinking about nature. Ethically speaking, this perspective begins with an interest in how flourishing may take place within a web of relationships, as opposed to one that begins with human rationality and the application of moral principles.

Another advantage of the relational approach is its effective inclusion of the temporal dimension. In contrast to many forms of art, environmental aestheticians argue that it is important to grasp the dynamic character of environment. This dynamic, changing character demands an ethic that recognizes not only evolving more-than-human ecologies, but also how human-nature narratives unfold over time and into the future in any particular landscape. Being sensitive to such narratives stands in some contrast to aesthetic preservationism, which seeks to preserve landscapes, often in their most natural state, and sometimes without acknowledging the role of, for example, indigenous communities, in shaping a landscape's environmental history (Holland and O'Neill 1998). As shown in our chapter on ecological restoration, working toward repairing anthropogenic damage requires an understanding of past, present, as well as future scenarios for the particular place being restored. In this way, a relational ethic more easily incorporates concerns of intergenerational justice, where ethical concerns meet aesthetic ones in terms of how to restore nature or design greenspaces, for example, that work toward harmonious coexistence of the human and more-than-human now and into the future. Beyond the cases we discuss in the book, art focusing on climate change brings home, via creative interventions, how anthropogenic carbon emissions have affected the atmosphere and what this means for present and future generations (Lippard 2007).

A relational approach also has the advantage of finding a role for disvalues in nature. On an intrinsic value approach, if a place is found to be ugly, for example, it may be harder to make the aesthetic preservationist's case stick.[2] In developing his work on environmental pragmatism, Light argues that benevolent environmental restoration practices can enable an experiential involvement that may help to restore a positive relationship with nature: "Restorationists get firsthand (rather than anecdotal and textbook) exposure to the actual consequences of human domination of nature. . . . Restoration is an obligation exercised in the interests of forming a positive community with nature" (Light 2000, 108). We do not mean to suggest that developing relationships will enable people to somehow begin to see such places as beautiful. Rather, as we mentioned early in the book, we think it is important to embrace the range of relationships and aesthetic values that occur through human-nature interactions. Besides obvious harmonious relationships, there will be other kinds of relationships that are meaningful and part of a flourishing life (at least in so far as we can know this to be the case for humans). Pointing out the limitations of intrinsic value approaches in environmental ethics, Holland makes the salient point that the "cheerful ascription of value does little to register the part played by the natural world in some of the grief, disappointment, sorrow and failure that are among the basic ingredients of a fully engaged, fully committed and fully vulnerable human life" (2012, 4–5). He argues for the importance of meaningful relations as a "unifying concept that characterizes evolutionary and ecological relations as well as cultural ones" (Holland 2012, 3). Bringing the concept of meaningful relations into our aesthetic and ethical reflections can enable us to recognize the significance of a range of our interactions with environments, how they can be enriching and educative, and how they might feed into ethical motivations. This provides a more complete and generous grasp of our meaningful engagement with environments in aesthetic terms, including, as it does, both positive and negative aesthetic relations, as well as those that mix the two.

Some of the ideas in Sheila Lintott's (2006) "eco-friendly aesthetics" serve to illustrate our thinking about aesthetics and ethics between nature and culture. Generally, eco-friendly aesthetics aims to address the negative consequences of aesthetic valuing on ecology, such as the taste for green lawns in some parts of the world. She discusses an especially interesting case for our purposes, the Mexican free-tailed bats residing in Congress Avenue Bridge in Austin, Texas. At first, Austin locals were generally unhappy about how the reconstruction of a city bridge created crevices that attracted thousands of the bats to roost in them. Many people find bats frightening and ugly, but the people of Austin came to appreciate the bats not only as the fascinating animals they are but also as aesthetically interesting, especially delighting in their roosting behavior every evening. Lintott does not explicitly discuss a relational approach, but her thoughts align with ours and help to show how our position differs from aesthetic preservationism. The local people, as well as visitors to the city, had the opportunity to come to know these very different creatures and their use of a particular urban place for habitat. The transition from vilifying the bats to marveling at them involved a process rooted in experiencing them on a daily basis as well as learning about the lives of the bats through an education campaign (Bat Conservation International 2017). Although we do not think that environmental education is always a prerequisite to developing pro-environment attitudes, nonetheless, Lintott's case study illustrates very well how firsthand experience and education enabled new people-to-bat attachments to form and a flourishing coexistence in the city.

The people of Austin came to appreciate the bats and their habits over time. Time is important in the development of relationships, for example, as we saw in the work of Lynne Hull, who developed an understanding of the needs of raptors in terms of safe places to roost and nest. Some of our examples in the book have focused on the role of the aesthetic practitioner who develops a relationship with nature through the creation of a garden, artwork, or stone wall. But we have also discussed the appreciative side where, for example, one enjoys the beauty of a well-formed hedge. One might object that our approach depends upon the development of relationships with nature over time, and that it does not leave room for the valuing that occurs in brief aesthetic encounters or in unforgettable "wow" moments. Striking moments such as these have a significant place in the range of aesthetic responses that people have, and we do not mean to discount their role in encouraging care and respect for environments. Aesthetic experiences within shorter time frames are potentially also strong motivators for protecting places, even if circumstances do not allow for deeper forms of attachment that may develop over time. The ecological artist comes to know her materials well and engages deeply with the site of the work, but part of that experience will involve many brief encounters that shape the relationship over time, and at any point along the way one may be said to develop a relationship through interaction. Relationships may be formed within shorter and longer time frames.

Another potential objection to a relational approach to how aesthetics supports ethics stems from the problem of how one can develop relationships with distant others. Some of the case studies discussed in the book depend upon firsthand experience or interaction with natural materials and processes, while others, such as some of the artworks discussed in chapter 5, depend upon a grasp of the work through photographs due to the artwork's ephemeral character or the remoteness of its site. One of the strengths of ecological art is its ability to bring to light both positive and conflicting relationships between humans and the rest of the natural world. Chris Jordan's photographic artworks convey the huge amounts of waste that humans produce, from an infinite mass of discarded cell phones to vast collections of plastic. His *Midway Project* took him to the Midway Atoll in the Pacific Ocean to photograph the effects of plastic pollution on Laysan albatrosses. In one photograph, "CF000313: Unaltered stomach contents of a Laysan Albatross fledgling, Midway Island, 2009," we discover direct connections between people in one part of the world and harm to distant nonhuman others thousands of miles away. These magnificent birds are seriously endangered by long-line fishing and plastic pollution, among other threats. As surface feeders, they swallow all kinds of plastic pollution floating on the surface of the ocean. Adult birds and chicks ingest the plastic, feel full, and die as a result of starvation and dehydration. This is evidenced in thousands of carcasses found on sites in the Pacific, where the world's plastic litter collects. There, as the birds decay, their remains expose what they swallowed that will not degrade for thousands of years: plastic lighters, small plastic toys, plastic water bottle tops. Jordan's work closes the gap, laying bare the ethical relationships between human and nonhuman on different sides of the world.

We have worked through some of the advantages of our aesthetic-ethical approach and addressed a few possible objections as well. In this concluding chapter, our aim has been to provide an alternative to aesthetic preservationism and one that is more suited to the kinds of human-nature interactions that take place beyond remote natural areas. We want to emphasize, again, that we do see a role for intrinsic valuing of nature and how such aesthetic valuing can help to motivate environmental protection. But we think it is important to register the broader range of relationships, especially in a world that is becoming more populated and developed. Our interest has been not in removing the cultural elements from aesthetic interactions with the natural world but rather to examine just what kinds of positive relationships might flow from them.

NOTES

1. Firth (2008) supports the position we take in the book by finding a place for aesthetics within Holland's narrative-based environmental ethics and its focus on meaningful relationships. In *Encountering Nature: Toward and Environmental Culture*, Heyd (2007) begins from the standpoint of humans and nature being intertwined, though his position does not explicitly foreground relationships in the way that we do. Rolston (2002) argues that beauty feeds into duty and points to the importance of ecological relationships. Saito's (2007) work on everyday aesthetics is notable for its discussion of both natural and cultural spaces, as well as its interest in ecology inspired by Aldo Leopold's work and the inclusive idea of "land as a community." We share both Rolston's and Saito's interests in ecological communities, but their approaches do not foreground the concept of relationship in the way that we do and, arguably, they favor a holistic approach that places value of the ecological whole above relationships between parts of that whole. See also Speed (2008).
2. Aesthetic preservationism aligns with "positive aesthetics," which we discussed in chapter 5. See also Brady (2013).

Bibliography

Aben, Rob, and Saskia de Wit. 1999. *The Enclosed Garden: History and Development of the* Hortus Conclusus *and Its Reintroduction into the Present-Day Urban Landscape*. Rotterdam: 010.

Andrews, Malcolm. 1999. *Landscape and Western Art*. New York: Oxford University Press.

Arntzen, Sven, and Emily Brady, eds. 2008. *Humans in the Land: The Aesthetics and Ethics of the Cultural Landscape*. Oslo: Oslo Academic Press/Unipub.

Attfield, Robin. 2001. "Postmodernism, Value and Objectivity." *Environmental Values* 10(2): 145–62.

Auping, Michael. 1983. "Earth Art: A Study in Environmental Politics." In *Humans in the Land: A Critical Anthology of Environmental Art*, edited by A. Sonfist, 92–104. New York: Dutton.

Bacon, Francis. 1950 [1625]. "On Gardening." Quoted in *The English Garden* by R. Dutton, 2nd ed. London: Batsford.

Baker, Steve. 2000. *The Postmodern Animal*. London: Reaktion.

Ballengée, Brandon. 2017. "Species Reclamation via a Non-Linear Genetic Timeline: An Attempted *Hymenochirus curtipes* Model Induced by Controlled Breeding." http://brandonballengee.com/species-reclamation. Accessed March 20, 2017.

Bat Conservation International, Congress Avenue Bridge. 2017. http://www.batcon.org/index.php/our-work/regions/usa-canada/protect-mega-populations/cab-intro Accessed June 30, 2017.

Beardsley, James. 1984. *Earthworks and Beyond*. New York: Abbeville Press.

Bearn, Gordon C. F. 1998. "Kitsch." In *Encyclopedia of Aesthetics*, edited by Michael Kelly, vol. 3. Oxford and New York: Oxford University Press.

Benson, John. 1999. "The Aesthetics of Stone-walling." *Thingmount Working Paper Series on the Philosophy of Conservation*, Lancaster University (manuscript).

Benson, John. 2008. "Aesthetic and Other Values in the Rural Landscape." *Environmental Values* 17(20): 221–38.

Berleant, Arnold. 1992. *The Aesthetics of Environment*. Philadelphia, PA: Temple University Press.

Berleant, Arnold. 1997. *Living in the Landscape: Towards an Aesthetics of Environment*. Lawrence: University Press of Kansas.

Berleant, Arnold, ed. 2002. *Environment and the Arts*. Aldershot: Ashgate.

Berleant, Arnold. 2004. *Re-Thinking Aesthetics: Rogue Essays on Aesthetics and the Arts*. Aldershot: Ashgate.

Berleant, Arnold, and Allen Carlson, eds. 2006. *The Aesthetics of Human Environments*. Peterborough: Broadview Press.

Bertelsen, Lars Kiel. 2004. "The Claude Glass: A Modern Metaphor between Word and Image." *Word and Image* 20(3): 182–90.

Boaz Simus, Jason. 2007. "A Response to Emily Brady's 'Aesthetic Regard for Nature in Environmental Art.'" *Ethics, Place and Environment* 10(3): 301–5.

Boaz Simus, Jason. 2008. "Environmental Art and Ecological Citizenship." *Environmental Ethics* 30(1): 21–36.

Boettger, Suzanne. 2002. *Earthworks: Art and the Landscape of the Sixties*. Berkeley: University of California Press.

Boetzkes, Amanda. 2010. *The Ethics of Earth Art*. Minneapolis: University of Minnesota Press.

von Bonsdorff, Pauline. 2005. "Agriculture, Aesthetic Appreciation and the Worlds of Nature." *Contemporary Aesthetics* 3. http://www.contempaesthetics.org/newvolume/pages/article.php?articleID=325&searchstr=bonsdorff. Accessed June 29, 2018

Boyer, M. Christine. 1994. *The City of Collective Memory: Its Historical Imagery and Architectural Entertainments*. Cambridge: MIT Press.

Brady, Emily. 2003. *Aesthetics of the Natural Environment*. Edinburgh: Edinburgh University Press.

Brady, Emily. 2006a. "The Aesthetics of Agricultural Landscapes and the Human-Nature Relationship." *Ethics, Place and Environment* 9(1): 1–19.

Brady, Emily. 2006b. "The Human-Nature Relationship in Environment and Land Art." In *Art, Ethics and Environment*, edited by A. Sigurjónsdóttir and Ó. Jónsson. Cambridge: Cambridge Scholars Press.

Brady, Emily. 2006c. "Aesthetics in Practice: Valuing the Natural World." *Environmental Values* 15: 277–91.

Brady, Emily. 2007. "Aesthetic Regard for Nature in Environmental Art." *Ethics, Place and Environment: Special Issue Environmental Art* 10(3): 287–300.

Brady, Emily. 2010. "Animals in Environmental Art: Relationship and Aesthetic Regard." *Journal of Visual Arts Practice* 9(1): 47–58.

Brady, Emily. 2013. *The Sublime in Modern Philosophy: Aesthetics, Ethics, and Nature*. Cambridge: Cambridge University Press.

Brady, Emily, and Isis Brook. 2003. "Topiary: Ethics and Aesthetics." *Ethics and the Environment* 8(1): 127–42.

Brady, Emily, and Pauline Phemister, eds. 2012. *Human-Environment Relations: Transformative Values in Theory and Practice*. Dordrecht: Springer.

Brady, Emily, and Jonathan Prior. 2017. "Environmental Aesthetics and Rewilding." *Environmental Values* 26: 31–51.

Brady, Emily, and Anita Seppä, eds. 1998. *IO: Journal of Applied Aesthetics Special Issue: Environmental Art*, 1. http://www.helsinki.fi/iiaa/io/io1998.pdf. Accessed July 11, 2018.

Breglia, Lisa C. 2006. *Monumental Ambivalence: The Politics of Heritage*. Austin: University of Texas Press.

Brook, Isis. 2003. "Making Here Like There: Place Attachment, Displacement and the Urge to Garden." *Ethics, Place and Environment* 6(3): 227–34.

Brook, Isis. 2005. "Can Merleau-Ponty's Notion of 'Flesh' Inform or Even Transform Environmental Thinking?" *Environmental Values* 14(3): 353–62.

Brook, Isis. 2010. "The Virtues of Gardening." In *Gardening: Philosophy for Everyone*, edited by O'Brien, D. Chichester: Wiley, 13–25.

Brook, Isis. 2012. "Make, Do, and Mend: Solving Placelessness Through Embodied Environmental Engagement." In *Human-Environment Relations: Transformative Values in Theory and Practice*, edited by Emily Brady and Pauline Phemister, 109–20. Dordrecht: Springer.

Brook, Isis. 2013. "Aesthetic Appreciation of Landscape." In *The Routledge Companion to Landscape Studies*, edited by Peter Howard, Ian Thompson, and Emma Waterton, 108–18. London: Routledge.

Brook, Isis. 2018. "Restoring or Restory-ing the Lake District: Applying Responsive Cohesion to a Current Problem Situation." *Environmental Values* 27(4): 427–45.

Brown, C., and Ted Toadvine, eds. 2003. *Eco-Phenomenology: Back to the Earth Itself.* New York: SUNY Press.

Budd, Malcolm. 2002. *The Aesthetic Appreciation of Nature*. Oxford: Oxford University Press.

Burke, Edmund. 1998 [1757]. *A Philosophical Enquiry into the Sublime and the Beautiful*. London: Penguin.

Carlson, Allen. 1986. "Is Environmental Art an Aesthetic Affront to Nature." *The Canadian Journal of Philosophy* 16(4): 634–50.

Carlson, Allen. 2000. *Aesthetics and the Environment: The Appreciation of Nature, Art and Architecture*. London and New York: Routledge.

Carlson, Allen. 2009. *Nature and Landscape: An Introduction to Environmental Aesthetics*. New York: Columbia University Press.

Carlson, Allen. 2010. "Contemporary Environmental Aesthetics and the Requirements of Environmentalism." *Environmental Values* 19(3): 289–314.

Carlson, Allen, and Sheila Lintott, eds. 2007. *Nature, Aesthetics, and Environmentalism: From Beauty to Duty*. New York: Columbia University Press.

Carter Park, Deborah, and Philip M. Coppack. 1994. "The Role of Rural Sentiment and Vernacular Landscapes in Contriving Sense of Place in the City's Countryside." *Geografiska Annaler B* 76(3): 161–72.

Castree, Noel. 2003. "Geographies of Nature in the Making." In *Handbook of Cultural Geography*, edited by Kay Anderson, Mona Domosh, Steve Pile, and Nigel Thrift. London: Sage.

CCBC. 1996. *History and Background Details to Penallta Community Park Millennium Project: Reclamation and Restoration of Penallta Tip and Nelson Bog.* Prepared by CCBC Client Services, June 14, 1996.

Chan, Elisabeth Clemence. 2009. "What Role for Ruins? Meaning and Narrative of Industrial Ruins in Contemporary Parks." *Journal of Landscape Architecture* 2: 20–31.

Clevely, Andi M. 1988. *Topiary*. London: Collins.

Clifford, Sue, and Angela King, eds. 1993. *Local Distinctiveness: Place, Particularity and Identity*. London: Common Ground.

Combe, William. 1878 [1812]. *Tour of Doctor Syntax in Search of the Picturesque.* London: Routledge & Sons.

Conesa, Hector M., Rainer Schulin, and Bernd Nowack. 2008. "Mining Landscape: A Cultural Tourist Opportunity or an Environmental Problem?" *Ecological Economics* 64(4): 690–700.

Cooke, Paul. June 2009. Development Officer, Groundwork Caerphilly, interview with Jonathan Prior.

Cooper, David E. 2006. *A Philosophy of Gardens*. Oxford: Oxford University Press.

Cooper, David E. 2012. *Convergence with Nature: A Daoist Perspective*. Totnes: Green Books.

Cooper, Paul. 2001. *Living Sculpture*. London: Mitchell Beazley.

Cosgrove, Denis. 1998. *Social Formation and Symbolic Landscape*. Madison: University of Wisconsin Press.

Costelloe, Timothy. 2013. *The British Aesthetic Tradition: From Shaftsbury to Wittgenstein*. Cambridge: Cambridge University Press.

Crawford, Donald. 1983. "Art and Nature: Some Dialectical Relationships." *Journal of Aesthetics and Art Criticism* 42(1): 49–58.

Crawford, Donald. 1993. "Comparing Natural and Artistic Beauty." In *Landscape, Natural Beauty and the Arts*, edited by Salim Kemal and Ivan Gaskell. Cambridge: Cambridge University Press.

Cronon, William. 1995. "The Trouble with Wilderness, or Getting Back to the Wrong Kind of Nature." In *Uncommon Ground: Toward Reinventing Nature*, edited by W. Cronon, 69–90. New York: Norton.

Crouch, David. 2010. "Flirting with Space: Thinking Landscape Relationally." *Cultural Geographies* 17(1): 5–18.

Cuomo, Chris, ed. 2003. *Ethics and the Environment Special Issue: Art*, 8(1).

Daniels, Neil. June 2009. Principal Landscape Architect, Caerphilly County Borough Council, interview with Jonathan Prior.

Daniels, Stephen. 1989. "Marxism, Culture, and the Duplicity of Landscape." In *New Models in Geography: The Political-Economy Perspective*, edited by Richard Peet and Nigel Thrift, 196–220. London: Unwin Hyman.

Davies, Stephen. 2012. *The Artful Species: Aesthetics, Art, and Evolution*. Oxford: Oxford University Press.

Denes, Agnes. 1998 [1982]. "Wheatfield: A Confrontation." In *Environmental and Land Art*, edited by Jeffrey Kastner and Brian Wallis. London: Phaidon.

Dezallier d'Argenville, Antoine Joseph. 1992 [1709]. "La Theorie et la Pratique du Jardinage." Quoted in *Gardens and the Picturesque* by John Dixon-Hunt. Cambridge: MIT Press.

Dia Art Foundation. 2017. "Walter DeMaria, *The Lightning Field*." www.lightningfield.org. Accessed March 20, 2017.

Drenthen, Martin. 2009. "Ecological Restoration and Place Attachment: Emplacing Non-Places?" *Environmental Values* 18: 285–312.

Dutton, Denis. 2010. *The Art Instinct: Beauty, Pleasure and Human Evolution*. Oxford: Oxford University Press.

Eaton, Marcia. 2001. *Merit: Aesthetic and Ethical*. Oxford: Oxford University Press.

Edensor, Tim. 2005. *Industrial Ruins: Space, Aesthetics and Materiality*. Oxford: Berg.

Egoz, Shelley, Jacky Bowring, and Harvey C. Perkins. 2001. "Tastes in Tension: Form, Function, and Meaning in New Zealand's Farmed Landscapes." *Landscape and Urban Planning* 57(3–4): 177–96.

Elliot, Robert. 1997. *Faking Nature: The Ethics of Environmental Restoration*. London: Routledge.

Farago, Jason. 2015. "Michael Heizer: 'I'm a Quiet Man. I Just Make Art." *Guardian*, May 18, 2015. https://www.theguardian.com/artanddesign/2015/may/18/michael-heizer-art- rocks-city. Accessed March 20, 2017.

Firth, Dan. 2008. "The Role of Aesthetic Considerations in a Narrative Based Approach to Nature Conservation." *Ethics and the Environment* 13(2): 77–100.

Fearnley-Whittingstall, Jane. 2002. *The Garden: An English Love Affair*. London: Weidenfeld Nicolson.

Fisher, John A. 2003. "Environmental Aesthetics." In *The Oxford Handbook of Aesthetics*, edited by Jerrold Levinson, 667–78. Oxford: Oxford University Press.

Fisher, John A. 2007. "Is It Worth It?: Lintott and Ethically Evaluating Environmental Art." *Ethics, Place and Environment* 10(3): 279–86.

Flam, Jack D., ed. 1996. *Robert Smithson: The Collected Writings*. Berkeley: University of California Press.

Foster, Jennifer. 2005. "Restoration of the Don Valley Brick Works: Whose Space?" *Journal of Urban Design* 10(3): 331–51.

Fudge, Erica. 2006. *Animal*. London: Reaktion.

Gablik, Susie. 2002. *The Reenchantment of Art*. London: Thames & Hudson.

Gabriel, Nate. 2011. "The Work That Parks Do: Towards an Urban Environmentality." *Social & Cultural Geography* 12(2): 123–40.

Giesecke, Annette, and Naomi Jacobs, eds. 2014. *The Good Gardener? Nature, Humanity and the Garden*. London: Artifice.

Gilpin, William. 1983 [1786]. "Northern Tour to the Lakes, etc. Made in 1772." Quoted in *The Picturesque* by Christopher Hussey. London: Frank Cass and Co.

Gilpin, William. 1792. *Three Essays*. London: R. Blamire.

Godlovitch, Stan. 1994. "Icebreakers: Environmentalism and Natural Aesthetics." *Journal of Applied Philosophy* 11(1): 15–30.

Godlovitch, Stan. 1998a. "Valuing Nature and the Autonomy of Natural Aesthetics." *British Journal of Aesthetics* 38(2): 180–97.

Godlovitch, Stan. 1998b. "Offending against Nature." *Environmental Values* 7(2): 131–50.

von Goethe, Johann Wolfgang. 1988. "Maxims and Reflections." In *Scientific Studies*, edited and translated by Douglas Miller. New York: Suhrkamp.

Goldsworthy, Andy. 1994. *Stone*. New York: Abrams.

Goldsworthy, Andy. 2000. *Wall*. New York: Abrams.

Goldsworthy, Andy, and Cumbria County Council. 2002. "Andy Goldsworthy Sheepfolds." Available at http://www.sheepfoldscumbria.co.uk/. March 20, 2017.

Grande, John K. 2003. *Balance: Art and Nature*. Montreal: Black Rose Books.

Grande, John K., and Edward Lucie-Smith. 2004. *Art Nature Dialogues: Interviews with Environmental Artists*. Albany: State University of New York Press.

Gray, Tyson-Lord. 2012. "Beauty or Bane: Advancing an Aesthetic Appreciation of Wind Turbine Farms." *Contemporary Aesthetics* 10. Available at http://hdl.handle.net/2027/spo.7523862.0010.011. Accessed May 23, 2017.

Groundwork Caerphilly. 1996. *Penallta Community Park Millennium Project: Development Strategy*. Groundwork Caerphilly.

Groundwork Islwyn and Rhymney. 1995. *Penallta Community Park: Millennium Bid Business Plan*. Groundwork Islwyn and Rhymney.

Gussow, Allen. 1972. *A Sense of Place: The Artist and the American Land*. San Francisco, CA: Friends of the Earth. Reprinted by Island Press, 1997.

Guyer, Paul. 2016. "18th Century German Aesthetics." In *The Stanford Encyclopedia of Philosophy*, Winter 2016 Edition, edited by Edward N. Zalta. Available at https://plato.stanford.edu/archives/win2016/entries/aesthetics-18th- german/. Accessed June 27, 2018.

Hall, Jeanette. 2015. "Rhododendron Ponticum and Its Invasive Hybrids." In *The Species Action Framework Handbook*, edited by Martin Gaywood, P. J. Boon, Desmond B. A. Thompson, and Ian Strachan, n.p. Scottish Natural Heritage. https://www.nature.scot/sites/default/files/Publication%202016%20-%20Species%20Action%20Framework%20Handbook%202016.pdf. Accessed July 11, 2018.

Hall, Matthew. 2011. *Plants as Persons: A Philosophical Botany*. Albany, NY: SUNY Press.

Hamilton, Andy. 2000. "The Art of Improvisation and the Aesthetics of Imperfection." *British Journal of Aesthetics* 40(1): 168–85.

Hargrove, Eugene. 1989. *Foundations of Environmental Ethics*. Denton: Environmental Ethics Books.

Harvey, David. 1996. *Justice, Nature and the Geography of Difference*. Oxford: Blackwell.

Hawkins, Harriet. 2012. "Geography and Art. An Expanding Field: Site, the Body and Practice." *Progress in Human Geography* 37(1): 52–71.

Hawkins, Harriet, and Elizabeth Straughan, eds. 2015. *Aesthetic Geographies*. Aldershot: Ashgate.

Haywood, Mark. 2006. "Claife Station: Looking at the World through Coloured Glasses." Paper presented at the Romantic Spectacle Conference, Roehampton.

Healy, Chris. 2001. "'Race Portraits' and Vernacular Possibilities: Heritage and Culture." In *Culture in Australia: Policies, Publics and Programs*, edited by Tony Bennett and David Carter, 278–98. Cambridge: Cambridge University Press.

Heartney, Eleanor. October 2003. "Mapping a Better World." *Art in America*, 91(10): 114–19.

Hepburn, Ronald. 1993. "Trivial and Serious in Aesthetic Appreciation of Nature." In *Landscape, Natural Beauty and the Arts*, edited by Salim Kemal, and Ivan, Gaskell, 65–80. Cambridge: Cambridge University Press.

Hepburn, Ronald. W. 2001. *The Reach of the Aesthetic: Collected Essays on Art and Nature*. Aldershot and Burlington: Ashgate.

Herrington, Susan. 2008. *On Landscapes*. New York: Routledge.

Hettinger, Ned. 2006. "Respecting Nature's Autonomy in Relationship with Humanity." In *Recognizing the Autonomy of Nature*, edited by Thomas Heyd. New York: Columbia University Press.

Hettinger, Ned. 2008. "Objectivity in Environmental Aesthetics and Protection of the Environment." In *Nature, Aesthetics, and Environmentalism: From Duty to Beauty*, edited by Allen Carlson and Sheila Lintott, 413–37, New York: Columbia University Press.

Heyd, Thomas. 2002. "Nature Restoration without Dissimulation: Learning from Japanese Gardens and Earthworks." *Essays in Philosophy* 2(1): 1–13.

Heyd, Thomas. 2006. "Thinking through Botanic Gardens." *Environmental Values* 15(2): 197–212.

Heyd, Thomas. 2007. *Encountering Nature: Toward an Environmental Culture*. Williston: Ashgate.

Higgs, Eric, Donald A. Falk, Anita Guerrini, Marcus Hall, Jim Harris, Richard J. Hobbs, Stephen T. Jackson, Jeanine M. Rhemtulla, and William Throop. 2014. "The Changing Role of History in Restoration Ecology." *Frontiers in Ecology and the Environment* 12(9), 499–506.

Hobbs, Richard J., and Jim Harris. 2001. "Restoration Ecology: Repairing the Earth's Ecosystems in the New Millennium." *Restoration Ecology* 9(2): 239–46.

Holland, Alan. 2012. "The Value Space of Meaningful Relations." In *Human-Environment Relations: Transformative Values in Theory and Practice*, edited by Emily Brady and Pauline Phemister, 3–15. Dordrecht: Springer.

Holland, Alan, and O'Neill, John. 1998. "The Integrity of Nature over Time." *Global Bioethics* 11(1–4): 9–18.

Hughes, Jon. December/January 2008. "What If Life Imitated Art?" *Ecologist*: 32–37.

Hull, Lynne. 2009. "The Mystery, the Evidence and the Small Atonements." In *Landviews*. http://www.landviews.org/articles/mystery-lh.html. Accessed March 20, 2017.

Hull, Lynne. 2017. "Artist Statement." http://greenmuseum.org/content/artist_index/artist_id-7.html. Accessed March 20, 2017.

Humphrey, Peter. 1985. "The Ethics of Earthworks." *Environmental Ethics* 7: 22–30.

Hunt, John Dixon. 1992. *Gardens and the Picturesque*. Cambridge, MA: The MIT Press.

Ingold, Tim. 2000. *The Perception of Environment: Essays on Livelihood, Dwelling and Skill*. New York and London: Routledge.

Instone, Lesley. 2001. *Lines Across the Land*. Monash: Monash University.

Instone, Lesley. 2004. "Situating Nature: On Doing Cultural Geographies of Australian Nature." *Australian Geographer* 35(2): 131–40.

Irwin, Robert. 1985. *Being and Circumstance: Notes toward a Conditional Art*. Culver City, CA: Lapis Press.

Jackson, Wes. 1992. "Nature as the Measure for a Sustainable Agriculture." In *Ecology, Economics, Ethics: The Broken Circle*, edited by F. Herbert Bormann and Stephen Kellert. New Haven, CT: Yale University Press.

James, Simon P. 2015. *Environmental Philosophy: An Introduction*. Cambridge: Polity.

James, Simon P. 2016. "The Trouble with Environmental Values." *Environmental Values* 25 (2016): 131–44.

Johnson, Kirk. 2012. "Christo's Colorado Project Delayed at Least till 2015." *New York Times*. February 22, 2012. http://www.nytimes.com/2012/02/23/us/christos-colorado-project-delayed-at- least-till-2015.html. Accessed March 20, 2017.

Jordan III, William R. 2003. *The Sunflower Forest: Ecological Restoration and the New Communion with Nature*. Berkeley: University of California Press.

Kant, Immanuel. 1963. *Lectures on Ethics* (1780–1781). Translated by Louis Infield. London: Harper and Row.

Kant, Immanuel. 2000. *Critique of the Power of Judgment*. Edited by Paul Guyer. Translated by Paul Guyer and Eric Mathews. Cambridge: Cambridge University Press.

Kastner John, and Brian Wallis, eds. 1998. *Environmental and Land Art*. London: Phaidon.

Katz, Eric. 1993. "Artefacts and Functions: A Note on the Value of Nature." *Environmental Values* 2(3): 223–32.

Katz, Eric. 2000. "The Big Lie: The Human Restoration of Nature." In *Environmental Restoration: Ethics, Theory, and Practice*, edited by W. Throop, 83–94. Amherst, NY: Humanity Books.

Kearns, Robin, Alun Joseph, and Graham Moon. 2010. "Memorialisation and Remembrance: On Strategic Forgetting and the Metamorphosis of Psychiatric Asylums into Sites for Tertiary Educational Provision." *Social & Cultural Geography* 11(8): 731–49.

Kelley, Caffyn. 2006. *Art and Survival: Patricia Johanson's Environmental Projects*. Salt Spring Island, BC: Islands Institute.

Kimbrell, Andrew. 2002. *Fatal Harvest: The Tragedy of Industrial Agriculture*. Washington, DC: Island Press.

Kimmelman, Michael. 2015. "Michael Heizer's Big Work and Long View." *New York Times*. https://www.nytimes.com/2015/05/17/arts/design/michael-heizers-big-work-and-long-view.html. Accessed March 20, 2017.

Kühn, Christian. 2005. "Complexity and Populism." In *What People Want*, edited by Michael Shamiyeh, 249–54. Dordrecht: Springer.

Lane, Andrew and Joyce Tait. 1990. *Practical Conservation: Woodlands*. London: Hodder and Stoughton.

Leddy, Tom. 2012. *The Extraordinary in the Ordinary: The Aesthetics of Everyday Life*. Peterborough: Broadview Press.

Levidow, Les, Pimbert, Michel, and Vanloqueren, Gaetan. September 2014. "Agro-ecological Research: Conforming or Transforming the Dominant Agro-Food Regime?" *Agroecology and Sustainable Food Systems* 38(10): 1127–55.

Lewis, Peter. June 2009. Project Manager, Caerphilly Countryside Service, Caerphilly County Borough Council, interview with Jonathan Prior.

Leopold, Aldo. 1989 [1949]. *A Sand County Almanac: And Sketches Here and There*. Oxford: Oxford University Press.

Light, Andrew. 2000. "Restoration or Domination?: A Reply to Katz." In *Environmental Restoration: Ethics, Theory, and Practice*, edited by W. Throop, 95–111. Amherst, NY: Humanity Books.

Light, Andrew. 2003. "The Urban Blindspot in Environmental Ethics." *Environmental Politics* 10(1): 7–35.

Light, Andrew, and Eric Katz, eds. 1996. *Environmental Pragmatism*. New York: Routledge.

Lintott, Sheila. 2006. "Toward Eco-Friendly Aesthetics." *Environmental Ethics* 28(1): 57–76.

Lintott, Sheila. 2007. "Ethically Evaluating Land Art: Is It Worth It?" *Ethics, Place and Environment: Special Issue: Environmental Art* 10(3): 263–77.

Lippard, Lucy. 2007. *Weather Report: Art and Climate Change*. Boulder, CO: Boulder Museum of Contemporary Art.

Littlewood, David. n.d. "Between Nature and Culture: The Place of the Garden in Narrative Approaches to Environmental Value." In *Thingmount Working Paper Series on the Philosophy of Conservation*. TWP Series 3–01, 1–23. Lancaster University. http://www.lancaster.ac.uk/users/philosophy/awaymave/onlineresources/thingmount.htm. Accessed July 11, 2018.

Lorimer, Jamie. 2015. *Wildlife in the Anthropocene: Conservation after Nature*. Minneapolis: University of Minnesota Press.

Mabey, Richard. 1980. *The Common Ground*. London: Hutchinson.

Mabey, Richard. 1996. *Flora Britannica*. London: Sinclair-Stevenson.

Mabey, Richard. 2000. *Country Matters*. London: Pimlico.

Marris, Emma. 2013. *Rambunctious Garden: Saving Nature in a Post-Wild World*. New York: Bloomsbury.

Maskit, Jonathan. 2007. "'Line of Wreckage': Towards a Post-Industrial Environmental Aesthetics." *Ethics, Place and Environment* 10(3): 323–37.

Matless, David, and George Revill. 1995. "A Solo Ecology: The Erratic Art of Andy Goldsworthy." *Cultural Geographies* 2: 423–48.

Maybee, Julie E. 2016. "Hegel's Dialectics." In *The Stanford Encyclopedia of Philosophy*, Winter 2016 Edition, edited by Edward N. Zalta. https://plato.stanford.edu/archives/win2016/entries/hegel-dialectics. Accessed July 11, 2018.

Mclean, Iain. 1997. "On Moles and the Habits of Birds: The Unpolitics of Aberfan." *Twentieth Century British History* 8(3): 285–309.

McKibben, Bill. 1989. *The End of Nature*. New York: Random House.

McNeill, Donald, and Mark Tewdwr-Jones. 2003. "Architecture, Banal Nationalism and Re-Territorialization." *International Journal of Urban and Regional Research* 27(3):738–743.

Mendieta, Ana. 1987. Unpublished statement. Quoted in J. Perreault, "Earth and Fire: Mendieta's Body of Work," in *Ana Mendieta: A Retrospective*. New York: New Museum of Contemporary Art.

Miller, Mara. 1993. *The Garden as Art*. Albany: State University of New York Press.

Monbiot, George. 2013. *Feral: Searching for Enchantment on the Frontiers of Rewilding*. London: Penguin.

Moore, Ronald. 2008. *Natural Beauty: A Theory of Aesthetics beyond the Arts*. Peterborough: Broadview Press.

Morris, Nina, and Sarah Cant. 2006. "Engaging with Place: Artists, Site-Specificity and the Hebden Bridge Sculpture Trail." *Social & Cultural Geography* 7(6), 863–88.

Morton, Timothy. 2007. *Ecology without Nature: Rethinking Environmental Aesthetics*. Cambridge, MA: Harvard University Press.

Nagel, Thomas. October 1974. "What Is It Like to Be a Bat?" *Philosophical Review* 83: 435–50.

Nicolson, Marjorie Hope. 1963. [1959]. *Mountain Gloom and Mountain Glory*. New York: Norton.

O'Neill, John, Alan Holland, and Andrew Light. 2007. *Environmental Values*. London: Routledge.

Orestano, Francesca. 2003. "The Revd William Gilpin and the Picturesque; Or, Who's Afraid of Doctor Syntax?" *Journal of Garden History* 31(2): 163–79.

Oxford English Dictionary. 2009. "regard, *n*." Third edition, December 2009; online version, March 2012. http://www.oed.com/view/Entry/161186. Accessed March 20, 2017.

Palmer, Clare, and Emily Brady. 2007. "Landscape and Value in the Work of Alfred Wainwright." *Landscape Research* 32(4): 397–421.

Parr, Hester. 2007. "Mental Health, Nature Work and Social Inclusion." *Environment and Planning D: Society and Space* 25(3): 537–61.

Parsons, Glenn. 2009. *Aesthetics and Nature*. London: Continuum.

Parsons, Glenn, and Allen Carlson. 2008. *Functional Beauty*. Oxford: Oxford University Press.

Patrizio, Andrew. 2011. *Stone: A Legacy and Inspiration for Art*. London: Black Dog.

Peterken, George. 1993. *Woodland Conservation and Management*, 2nd ed. London: Chapman and Hall.

Peterson, Anna. 1999. "Environmental Values and the Social Construction of Nature." *Environmental Ethics* 21(4): 339–57.

Petherick, Tom. 2010. *Biodynamics in Practice*. Forest Row: Sophia Books.

Petts, Mick, and B. Dale. 2000. *Penallta Community Park: Artworks Strategy*. Produced for Groundwork Caerphilly, March 21, 2000.

Phillips, G. 2003. *Groundwork Caerphilly: 12 Month Residency Final Report*. Caerphilly: Groundwork Caerphilly.

Pollan, Michael. 2006. *The Omnivore's Dilemma*. London: Bloomsbury.

Pope, Alexander. 2002 [1731]. "Epistle to Lord Burlington." Quoted in *The Story of Gardening* by Penelope Hobhouse. London: Dorling Kindersley.

Potteiger, Matthew, and Jamie Purinton. 1998. *Landscape Narratives: Design Practices for Telling Stories*. New York: John Wiley & Sons.

Price, Uvedale. 1842 [1794]. *On the Picturesque.* Edinburgh: Caldwell, Lloyd and Co.

Prior, Jonathan. 2012. "The Roles of Aesthetic Value in Ecological Restoration: Cases from the United Kingdom." PhD diss., The University of Edinburgh.

Prior, Jonathan, and Emily Brady. 2017. "Environmental Aesthetics and Rewilding." *Environmental Values* 26(1): 31–51.

Prum, Richard. 2017. *The Evolution of Beauty*, New York: Doubleday.

Qviström, Mattias. 2007. "Landscapes out of Order: Studying the Inner Urban Fringe beyond the Rural—Urban Divide." *Geografiska Annaler: Series B, Human Geography* 89(3): 269–82.

Rackham, Oliver. 1977. *Trees and Woodland in the British Landscape*. London: Dent.

Rags Over Arkansas River (ROAR). 2016. http://www.roarcolorado.org/about-us.html. Accessed April 19, 2016.

Regan, Tom. 1991. "The Case for Animal Rights." In *Animal Experimentation*, edited by Robert Baird and Stuart Rosenbaum, 77–88. London: Prometheus Books.

Repton, Humphrey. 1842 [1794]. "Letter to Uvedale Price, Esq." In *On the Picturesque* by Uvedale Price, 411–16. Edinburgh: Caldwell, Lloyd and Co.

Rhemtulla, Jeanine M., and William Throop. 2014. "The Changing Role of History in Restoration Ecology." *Frontiers in Ecology and the Environment* 12: 499–506.

Riedelsheimer, Thomas (Dir.). 2004. *Andy Goldsworthy: Rivers and Tides: Working with Time*. DVD film; Prod. A. Donop.

Rolston III, Holmes. 2002. "From Beauty to Duty: Aesthetics of Nature and Environmental Ethics." In *Environment and the Arts: Perspectives on Environmental Aesthetics*, edited by Arnold Berleant, 127–41. Aldershot and Burlington: Ashgate.

Rosenthal, Michael. 1983. "Some Attitudes of Earth Art: From Competition to Adoration." In *Humans in the Land: A Critical Anthology of Environmental Art*, 92–104. New York: Dutton.

Ross, Stephanie. 1993. "Gardens, Earthworks, and Environmental Art." In *Landscape, Natural Beauty and the Arts*, edited by Salim Kemal and Ivan Gaskell, 158–82. Cambridge: Cambridge University Press.

Ross, Stephanie. 1998. *What Gardens Mean*. Chicago, IL: Chicago University Press.

Ross, Stephanie. 2006. "Paradoxes and Puzzles: Appreciating Gardens and Urban Nature." *Contemporary Aesthetics* 4.

Ruiz-Ballesteros, E., Valcuende, J. M., Quintero, V., Cortes, J. A. and Rubio, E. 2009. "Naturalizing the Environment: Perceptual Frames, Senses and Resistance." *Journal of Material Culture* 14(2): 147–67.

Rynszteijn, Madeleine. 2002. *Olafur Eliasson*. London: Phaidon.

Sadler, R. 1993. *Public Perceptions of Short Rotation Coppice.* London: UK Department of Trade and Industry Report.

Saito, Yuriko. 1992. "The Japanese Love of Nature: A Paradox." *Landscape* 31: 1–8.

Saito, Yuriko. 1996. "Japanese Gardens: The Art of Improving Nature." *Chanoyu Quarterly: Tea and the Arts of Japan* 83: 40–61.

Saito, Yuriko. 1997. "The Japanese Aesthetics of Imperfection and Insufficiency." *Journal of Aesthetics and Art Criticism* 55(4): 377–85.

Saito, Yuriko. 1998a. “Appreciating Nature on Its Own Terms.” *Environmental Ethics* 20(2): 135–149.

Saito, Yuriko. 1998b. “The Aesthetics of Unscenic Nature.” *Journal of Aesthetics and Art Criticism* 56(2): 101–11.

Saito, Yuriko. 1998c. “Japanese Aesthetics: Historical Overview.” In *Encyclopedia of Aesthetics, volume 2*, edited by Michael Kelly, 547–50. Oxford and New York: Oxford University Press.

Saito, Yuriko. 2002. “Environmental Directions for Aesthetics and the Arts.” In *Environment and the Arts*, edited by Arnold Berleant. Aldershot: Ashgate.

Saito, Yuriko. 2007. *Everyday Aesthetics*. Oxford: Oxford University Press.

Saito, Yuriko. 2017. *Aesthetics of the Familiar: Everyday Life and World-Making*. Oxford: Oxford University Press.

Sauer, Carl. 1925. *The Morphology of Landscape*. Berkeley, CA: University of California.

Scarbrough, Elizabeth. 2014. “Unimagined Beauty.” *Journal of Aesthetics and Art Criticism* 72: 445–49.

Schauman, Sally. 1998. “The Garden and the Red Barn: The Pervasive Pastoral and Its Environmental Consequences.” *Journal of Aesthetics and Art Criticism, Special Issue: Environmental Aesthetics* 56(2): 181–90.

Serra, Richard, and Dirk Reinartz. 1995. *Afangar*. Zurich: Scalo.

Shapiro, Gary. 1997. *Earthwards: Robert Smithson and Art after Babel*. Berkeley: University of California Press.

Shiva, Vandana. 2016. *Who Really Feeds the World? The Failures of Agribusiness and the Promise of Agroecology.* Berkeley, CA: North Atlantic Books.

Sigurjónsdóttir, Aesa, and Ólafur Páll Jónsson, eds. 2006. *Art, Ethics and Environment.* Cambridge: Cambridge Scholars Press.

Smithson, Robert. 2005. *Floating Island*. www.robertsmithson.com/sculpture/floating_island.htm. Accessed March 20, 2017.

Snow, Dan. 2001. *In the Company of Stone. The Art of the Stone Wall*. New York: Abrams.

Sonfist, Alan, ed. 1983. *Art in the Land: A Critical Anthology of Environmental Art*. New York: Dutton.

Soper, Kate. 2003. “Privileged Gazes and Ordinary Affections: Reflections on the Politics of Landscape and the Scope of the Nature Aesthetic.” In *Deterritorialisations: Revisioning Landscapes and Politics*, edited by Mark Dorrian and Gillian Rose. London: Black Dog, 338–48.

Spaid, Sue. 2002. *Ecovention: Current Art to Transform Ecologies*. Greenmuseum.org; The Contemporary Arts Centre; ecoartspace.

Speed, Fran. 2008. “Meaningful Relations: The Relational Basis of Aesthetic Judgement and Its Implications for Aesthetic Concern in Environmental Deliberation.” PhD diss., University of Central Lancashire.

Spence, Mark David. 2000. *Dispossessing the Wilderness. Indian Removal and the Making of the National Parks*. New York: Oxford University Press.

Stead, Naomi. 2004. “The Semblance of Populism: The National Museum of Australia.” *Journal of Architecture* 9(3): 385–96.

Steenbergen, Clemens, and Wouter Reh. 1996. *Architecture and Landscape.* Bussum: THOTH.

Switzer, Stephen. 2001 [1718]. "Iconographia Rustica." Quoted in *The English Garden: A Social History* by Charles Quest-Ritson. London: Penguin.

Taylor, Paul. 1996. *Respect for Nature*. Princeton, NJ: Princeton University Press.

Thompson, Janna. 1995. "Aesthetics and the Value of Nature." *Environmental Ethics* 17(3): 291–305.

Thompson, Paul B. 2017. *The Spirit of the Soil: Agriculture and Environmental Ethics*, 2nd ed. New York: Routledge.

Throop, William. 2000. "Introduction." In *Ecological Restoration: Ethics, Theory, and Practice*, edited by William Throop, 11–23. Amherst: Humanity Books.

Tomkins, Carl. 2003. "Flying into the Light." *New Yorker*, January 13, 2003.

Townsend, Dabney. 1997. "The Picturesque." *Journal of Aesthetics and Art Criticism* 55(4): 365–76.

Tuan, Yi-Fu. 2004. *Dominance and Affection: The Making of Pets*. London and New Haven: Yale University Press.

Tufnell, Ben. 2006. *Land Art*. London: Tate.

Turner, Frederick. 1990. "The Self-effacing Art: Restoration as Imitation of Nature." In *Restoration Ecology: A Synthetic Approach to Ecological Research*, edited by William R. Jordan III and Michael E. Gilpin, 47–50. Cambridge: Cambridge University Press.

United Nations, Department of Economic and Social Affairs, Population Division. 2015. World Urbanization Prospects: The 2014 Revision (ST/ESA/SER.A/366).

van Diggelen, Rudy, Albert P. Grootjans, and Jim Harris, 2001. "Ecological Restoration: State of the Art or State of the Science?" *Restoration Ecology* 9(2): 115–8.

Vogel, Steven. 1996. *Against Nature*. Buffalo: State University of New York Press.

Vogel, Steven. 2015. *Thinking Like a Mall: Environmental Philosophy after the End of Nature*. Cambridge: MIT Press.

Walsh, Kevin. 1992. *The Representation of the Past: Museums and Heritage in the Post-Modern World.* London: Routledge.

Walton, Kendall. 1970. "Categories of Art." *Philosophical Review* 79 (3): 334–67.

Watelet, Claude-Henri. 2003. [1774]. *Essay on Gardens: A Chapter in the French Picturesque*, edited and translated by Samuel Danon. Philadelphia: University of Pennsylvania Press.

Whatmore, Sarah. 2002. *Hybrid Geographies: Natures Cultures Spaces.* London: Sage.

White, Peter S., and Joan L. Walker. 1997. "Approximating Nature's Variation: Selecting and Using Reference Information in Restoration Ecology." *Restoration Ecology* 5(4): 338–49.

Winkler, Justin. 2005. "The Eye and the Hand: Professional Sensitivity and the Idea of an Aesthetics of Work on the Land." *Contemporary Aesthetics* 3. www.contempaes.org. Accessed March 20, 2017.

Winlock, Herbert Eustis. 1942. *Excavations at Deir el Bahri 1911–31.* New York: Macmillan.

Winter, Deborah du Nann, and Sue Koger. 2003. *The Psychology of Environmental Problems.* Lawrence, KS: Erlbaum Associates.

Worster, Donald. 1999. "Good Farming and the Public Good." In *Environmental Ethics: Concepts, Policy, Theory*, edited by J. DesJardins. Mountain View, CA: Mayfield.

Wu, Xin. 2013. *Patricia Johanson and the Re-Invention of Public Environmental Art, 1968–2010*. Farnham: Ashgate.

Yusoff, Kathryn, and Jennifer Gabrys. 2006. "Time Lapses: Robert Smithson's Mobile Landscapes." *Cultural Geographies* 32: 444–50.

Zimmerer, Karl S. 2000. "The Reworking of Conservation Geographies: Nonequilibrium Landscapes and Nature-Society Hybrids." *Annals of the Association of American Geographers* 90(2): 356–69.

Index

About the Authors

Emily Brady is professor of philosophy and Susanne M. and Melbern G. Glasscock chair and director of the Glasscock Center for Humanities Research at Texas A&M University. She works in the areas of aesthetics and philosophy of art, environmental ethics, environmental humanities, and animal studies. She has authored or coedited six books, most recently, *The Sublime in Modern Philosophy: Aesthetics, Ethics and Nature* (2013).

Isis Brook is a philosopher who has published on landscape aesthetics and has a research interest in our responses to gardens and landscapes, and how these responses can be made more sensitive and ultimately shape who we are. She is head of faculty at Crossfields Institute. Previously, she held philosophy lectureships at Lancaster University, where she also served as faculty teaching dean, and the University of Central Lancashire.

Jonathan Prior is lecturer of human geography at Cardiff University, whose research interests straddle environmental aesthetics, landscape policy, and sound studies. In particular, he is interested in the ways in which humans modify landscapes, particularly through environmental conservation and ecological restoration, and how these modifications intersect with different types of environmental and cultural values.

www.ingramcontent.com/pod-product-compliance
Ingram Content Group UK Ltd.
Pitfield, Milton Keynes, MK11 3LW, UK
UKHW012042150926
475560UK00001B/142

9 781786 610751